THE EINSTEIN FABLES

by

James Conor O'Brien

Front cover piece Flammarion Engraving

This work is a body of fiction parodying the biographies and lives of scientists and artists, and is not intended in any way to represent the actual lives of those portrayed, except the bit about Tycho Brahe's moose.

To
Eric M. Rogers
who awoke the kraken of physics within me

Newton – A Study in Infrared

For Newton had a weakness, and it was Women.

Newton was a tall fierce looking man who often turned heads, with his splendid profile and insightful remarks one might have assumed upon first meeting this distinguished gentlemen he might be the very apex of a ladies man, but alas, time and the ingestion of chemical Mercury had done to his mind what the wind does to a dandelion clock – it had quite blown away; and while in his later years Newton remained a formidable calculating machine, in many respects he was as daft as a windsock in a winter storm.

What was most peculiar about his nature was his singular ineptness with the female form, mind and manners.

"Master, a woman is here to see you," Newton's manservant Hedby knocked on the door to his study.

"Is there now," said Newton absent-mindedly, "bring him in."

"His a Her, my lord."

"Yes, in here."

"No, my lord, I said Her not here?"

"That's what I heard you say man," Newton looked up from his journal sharply. "Did you say a Her? You mean a her-man?"

"Yes, my lord as in one-of-them."

"Send for the bailiff," he whispered as he rose; making a dash for the garden door, "- and have the creature escorted out of here."

It was, however, too late as she appeared at the door and entered as a social juggernaut.

"Ah, there you are Sir Isaac," a woman dressed in as little apparel as the age would allow, wafted into the study offering him her hand, "I am so very pleased to make your acquaintance."

"Please come in," he stared at the hand and after a moments hesitation held it limply and then pointed vaguely about the room.

"How kind of you," she sat on a chair by the fireplace and like a cat watched him for about a minute then decided to help him."I take sugar."

"Hedby, bring some sugar, for …I'm sorry I didn't catch your name."

"Lady Ramsgate, wife of the previous Lord Ramsgate. I sent you a letter; actually I've sent you about nine letters. I thought I'd pop in to see if you have received them."

"Aha! That Lady Ramsgate, yes of course," he leaned against the first place and nudged the burnt remains of the letters beneath the firedogs. "I've been expecting you, and to what matter do I owe this pleasure?"

"Didn't you read my letters?"

"Well, my Lady so many affairs of the college and the Mint and so forth, I don't suppose you could remind me?"

"Hmm," she lowed her head and frowned at him. "You will remember that the late Lord Ramsgate …"

At the moment Hedby returned with a small sack of Indian sugar. "Sugar, my Lady."

"Yes," Newton threw up his hands, "of course you wrote to me concerning sugar, why here Lady Ramsgate some sugar for you. It's been so good of you to visit."

She looked at them both for a moment then accepted the nature of the Cambridge don as not being a polite fiction.

"How nice," she smiled, "I don't suppose you have any tea to go with it?"

"Tea! Hedby!"

"Tea?"

"Yes, damn your eyes, tea!" He flustered. "As in a beverage

brewed from the leaves of the herb Sinus Camillus! Didn't that Lady Catherine of Braganza send us some?"

"At once my lord."

"Now then," Newton put his hands together and smiled wanly, "where were we?"

"Sir Isaac, I see," she began as she sucked on a cube of sugar, "... I say this is rather good sugar. Where did you get it?"

"From Jamaica, my Lady, I use it in my alchemical studies."

"Oh well," she dropped it on the table, "never mind. The point is, as you know, the recent death of my husband was unusual."

"Unusual? In what way madam?"

"He was murdered."

"Most unusual."

"Quite, now the thing is..."

"Tea, my Lady," Hedby returned carrying a small jar of tea which he proffered.

"Aren't you going to put it hot water?"

"Hedby! Water!"

"Sir Isaac," lady Ramsgate leaned her chin on her hand and stared at Isaac as Hedby disappeared once more, "am I correct in assuming that you do not get many ladies visiting?"

"Not if I ... err, yes quite correct, my lady, so much work and what not."

"How many precisely?"

"Including yourself, my lady?"

"If you wish."

"Let me see, possibly, there was the Lady Catherine of Braganza."

"You mean the Queen, when she came and knighted you?"

"Got married did she?"

"Possibly - is she not a lady?"

"Portuguese." Newton pointed out.

"Ah," she smiled, "Anyone else?"

"Including yourself."

"If you wish."

"Well, I make that about two. Yes, that would be about right, two, possibly three, but only because the chancellor wears a long cape."

"Great Heavens! What is wrong with you?"

"Madam, I assure there is nothing wrong with me, it is simply the rules of the college. Cambridge is, after all, a university of learning and wisdom, there simply is no necessity to meet women here, and these rules have been handed down to us since it's inception. Which in so doing, we are carrying on the great scholastic tradition of education and scholarship, that has made Cambridge renowned throughout the enlightened world. "

"So, you're all afraid of women, is that it?"

"A little," he blushed.

"Very well, that is no concern of mine - and will I never get to explain my reason for coming here?"

"Water, my lady?" Hedby returned, holding a kettle and wooden dish.

"Very good, now please put them all together before I start screaming."

"Hurry up, Hedby," Newton shrieked, "we don't want a screamer!"

"Sir Newton, or may I call you Isaac?" Lady Ramsgate turned her long neck and placated him with a smile. "I was being rhetorical about screaming, there's really no need to vex."

He looked her in the same way as a hyena may look up a raging bull elephant - with cackling dread.

"Well, to continue," she said, "As you may well know, my

husband was murdered most foully last year."

"Was he run over by a wagon filled with poultry?" asked Hedby as he mixed a full bag of sugar, tea and tepid water in the wooden bowl.

"Not quite," she looked at the manservant with some annoyance, "he was stabbed."

"By a duck or a chicken?" asked Hedby as he offered up the bowl.

"I think you need to get off this idea concerning fowls – and do run and get me a cup."

"A cup!" cried Newton, whose backbone had now completely collapsed in the presence of lady Ramsgate's serene poise.

"Yes, so to bring this discussion forward, if that is possible, I have upon decided learning of your position of Warden of the Mint where you have expedited the trial numerous ruffians to justice, to engage you in this matter of the investigation of Arthur's death."

"This is quite good," said Newton as he sipped the bowl, "and you call it tea?"

"I do wish you would pay attention. So, will you agree to investigate my husband's death?"

"Ah, yes but of course, my lady, will you be staying for biscuit?"

"Aren't you going to ask me some questions?"

"Type of biscuit? I think we have ginger snaps."

"Such as," she continued with the unflappable insistence of a dodo determined on extinction, "whom might have motive, whom do I suspect, did Arthur have any enemies? Those sort of questions."

"Will it help?"

"Really, Sir Newton, your reputation does not do you justice."

"My lady," interposed Hedby with a strange smile, "if I may speak my lady. His lordship has his methods, which may seem unusual from lack of familiarity, but nevertheless, you may rest assured, he will endeavour in this matter and bring the question of your husband's death to a resolution."

Newton looked startled.

"Really?" she looked at Hedby with new appraisal. "Well, in that case I will leave the matter in your capable hands, Sir Newton. It has so … so interesting meeting you. I must be off."

She stood up and looked him full in the face.

"Oh good," said Newton, as his eyes darted about the room trying to find other eye sockets to hide in, "I mean yes."

"It would mean so much to me to find out the truth of this awful matter. Goodbye."

"Farewell, lady Ramsgate."

There was an awkward moment as she offered her hand to Newton, who looked at with the perplexed gaze of mongoose trying to understand the workings of clock. To which she sighed, and grandly slipped out the door as if sliding across ice.

"Is she gone?" Newton piped. "Damn your eyes Hedby! Why did you have to agree for me? I almost had gotten rid of her and now we have to help that ghastly harpy!"

"I thought she was quite lovely, my lord."

"That's besides the point, now I have to meet her again, in person. She's a woman you know?"

"I had noticed, my lord."

"I didn't become a physicist for the wild parties! Blast!"

Thus, it was, that Isaac Newton creator of the New Sciences and Warden of the Mint was engaged to investigate a murder. The main problem, of course, was not solving a crime, but rather the dread of having to meet, at some point in the future, with that woman again.

Before the advent of street lighting, the city of London at night seemed a jagged landscape of giant boulders and soaring plinths, lit by the occasional flare of an door opening or call of a drunken reveller, awash with the endless showers, the smell of decay and the sounds of echoing streets. A strange hostile place, quite distinct from the day filled with carts trundling past, screaming children and the comforting call of the town crier, the nightscape of London was a place of knives and black deals.

As Newton sheltered in the shadow of the building, the sharp features of his white face glowing like a beacon in the murk, he watched the windows on the building opposite with the intensity of an eagle searching for prey. So intent, that he almost fell over when the nightwatchman tapped him on the shoulder.

"'Ere you, what in blazes are you doing?"

"Newton! Warden of the Mint!"

"Who? What?"

"I am Sir Isaac Newton, Warden of the Mint. I am here on official business."

"Never heard of you," the watchman barked, "This your street? Where do you live?"

"Look my fellow, I am the Newton, don of Cambridge, knighted by her majesty and renown throughout the world of letters and natural philosophy."

"And I'm Jock, keeper of this lantern and holder of this street. And you matey, are nicked!"

"You can't arrest me, I'm an officer of the crown, a warden of the realm. I'm your superior!"

"Where's your lantern?"

"What?"

"All watchmen got to have a lantern. Them's the regulations, let's see ya lantern."

"I don't have one, I'm incognito, and I'm spying on a villain."

"A spy! Right matey! Now you're truly nicked!"

The watchman grabbed Newton by the collar and hauled him down the road, all the while Newton protesting not only his innocence but also his royally appointed position.

From across the street a window opened a fraction further as a sinister face leaned out the window and with a smile watched the retreating pair, then behind him a light flared up and the strange face in the window quickly disappeared. This was followed by shrieks of laughter.

Newton was thrown into Newgate, the debtor's prison, as that night the cities actual gaol was full with paying customers. The captain of the night watch was a man of considerable financial cunning and had recently discovered that carters from out of London would pay a bright penny to sleep with their carts inside the secure walls of the gaol, as this was far safer than risking the ale houses or boarding rooms outside. Hence, the debtors prison took care of all the real prisoners; the side affect from this was that people were often lost within the debtors prison until they managed to pay their way out. Since, the captain of the watch ran both the goal and the debtors prison, he was able to make quite a few bright pennies.

"Wot you in fur matey?" a mouthful of black and broken teeth breathed into Newton's ear, with a breath that could have corroded platinum.

"Keep off me, you villain! I have done nothing, I am innocent!" Newton declared as he jumped away from his fellow prisoner.

"Heard that one, afore," grinned his toothy interlocutor.

"Confound you fiend, there has been a mistake. I am Sir Isaac Newton, Warden of the Mint and president of the Royal Society."

"Maybe you should talk to Pope over there with his mate the King of France. Right nut house this place is turning into."

"Look I really am Isaac Newton, I am investigating a murder…"

"What the bleedin' 'eck is Sir Isaac Bleedin' Newton doing investigating a murder?"

"Well I…"

"I mean, it just doesn't stand to reason and all, on the one hand you have that Mister Newton who did all them scientific things, and on the other we got you, some ponce with a cape all locked up with a hundred other buggers in Fleet Prison with myself, Harry Norton reknown throat cutter and buggerer. I mean, if you consider the balance of the probabilities it's pretty bloody unlikely you're Sir Isaac Bleedin' Newton. Ain't it?"

"Since you put it like that," Newton coughed and kept his back against the wall. "A case of mistaken identity."

"Aye, so hand over ya purse so I can buy me way out or I'll remove your ability to reason."

"Buy your way out?"

"It's a debtors prison matey, you stay in here till you pay the guard for unpaid bills, how else do you expect to get out?"

"Guard!"

A short while later Newton found himself out on the street, somehow not only buying his own way but Master Harry Norton's freedom as well.

"Well, matey I'll be thanking ya for helping me out of that clink, ah, the sweet taste of freedom."

"You had a knife at my back when I was talking to the guard, I didn't really have any choice."

"Oh no matey, that wasn't a knife, I was just being friendly like. Right then, matey, buy me a drink and we'll call it square."

"Square? How does that square us off? That is neither not even triangular nor circular."

"Er'e now, I told you how to get out and all, didn't I? You'd still be in there with all them's criminals and buggers. Come on, show some appreciation."

"As long as you don't stick a knife in my back."

"Well, maybe not your back."

The tavern, Harry lead them to, was more an underground sewer than what could be comfortably called an establishment, one had only to stand at the bench that served as a bar to watch the dirt float past to realise how deep beneath the streets of London they had descended. Large pyramids of refuse and broken bottles stood about the room serving as ad hoc walls and partitions to rooms and fireplaces, as groups of men and women drank in their natural habitat, screaming with abandon.

Newton stared as his drink on the bar and watched in amazement as a small rat crawled out of the mug, curled up in a ball on the bench and died. He pushed the mug to one side.

"Look, Harry, I would like to ask you some questions."

"Sure matey," as he pulled his own rat out of his cup. "Corse, if you're looking for information, that'll cost. Business is business, even down here."

"I'd swear you'd sell your own grandmother."

"I got a better price for me mum, wife and daughters," Harry grinned broadly, "I like to think big, and no one wanted me gran."

"Yes, well, such things aside, I'm investigating the death of Lord Ramsgate. He was slain not too far from here, by unknown assailants. I need to find out their names and where they reside."

"Ah, do you now? Hmm, lets see ... who killed Arthur?"

"Arthur? How did you know his name?"

"Who Arthur? Well everyone knows Arthur, right regular

down here he is, saw him couple days ago, I did."

"No you're mistaken, Lord Ramsgate died several weeks ago."

"Lord Arthur Ramsgate?"

"None other."

"Well matey, I ain't mistaken then, he was he're day afore last, that's for sure."

"But that's impossible.

"I wouldn't know about that', matey, but I do know I saw his lordship day before last. That'll cost ya a couple of bob."

"Bob? Whose Bob?"

"Shillings, ya daft git."

"Damn your eyes, here," Newton turned his back and prised out the token coins as Harry looked over his shoulder. "So, you definitely saw him but that would mean… it must mean that Lord Arthur Ramsgate is still alive."

"You're fast one. Teach that at college did they?"

"But why, why was he reported dead?"

"Who says he's dead?"

Newton looked closely at Harry and had a flash of inspiration.

"That'll will cost you a couple of bob."

"Eh?"

"Information costs."

"Yeah but…"

"You want me to ask you questions and pay you."

"Sure but…"

"So to do that, you have to make sure I ask the right questions, and to find that out you have to ask me questions to know what I really want."

"Ah."

"Two bob."

"I see now why you're a lordship," Harry mumbled as he slid

over the two shillings. "So, who says his dead?"

"Lady Ramsgate."

Newton then stood poised like a mongoose hypnotising a snake, the silence grew into a moment of tension until Harry broke.

"And you want me to ask, where I think Lord Ramsgate is?"

"Two bob."

"Haggling ain't what it used to be." Harry looked bemused, "So where do I think his lordship is?"

"You will have to tell me."

"I .. I think Arthur is shacked up with some tart in Hanging Sword Alley. Did I just say that?"

"Two bob. Thank you."

"No problem matey, just remind me to never play cards with you."

"Come, I need your assistance in a matter of some indiscretion."

"Will this cost me? No wait, don't answer that."

Another short while later, the two improbable pair stood in the darkness before the same building Newton had been arrested originally.

"This is where I was told Lord Ramsgate was murdered. It is most curious that this is the same place you say he is still residing."

"I'd say something, but I'm worried it will lead me back to the debtor's prison."

"Right, Harry, since you are a hoodlum…"

"Them's me credentials," said Harry with a grin.

"You must enter that building and find out who is living within. I will keep watch and arrest you when you return."

"Eh?"

"You have a question?"

"Err," Harry pulled his earlobe and looked at Sir Isaac with new insight. "You're going to pay me, right?"

"Yes, of course. What is your fee?" Handing over two shillings.

"Well, it will be a bit more than this, I'll tell you." Then shrugged. "Ah, what the hell, something's bound to turn up."

"Good man. Do it quickly and there will be a bonus."

"Well, in that case. I'll be back before you can say Harry Norton!"

"Harry Norton, I arrest you in the name of the crown."

"What?"

"Joking. Now get on with it man."

Harry vanished into the shadows, while Newton's pale face loomed once more in approaching dawn. There was a tinkle of glass followed by a shout, a scream and the clatter of feet.

"Bleedin 'ell!!" cried the erstwhile thief as he crashed down the road pursued by a couple of border terriers. "Hounds!"

Newton strode across the street and confronted a man dressed only in a nightshirt.

"Lord Ramsgate, so good of you to meet me."

"Blast. You're from the wife aren't you?"

"That will cost you two shillings." Newton grinned.

Einstein and his Muse

As a young man Einstein was not always clever - Einstein

One summer's day in 1897 on a dusty byroad, a pimply and dishevelled Albert Einstein was lost in the Florentine countryside. With the finish of his school exams his parents had armed him with only a bicycle, a backpack of bagels, and a map left over from the Napoleonic Wars. Then sent him over the mountains from Geneva, and he was now spending time discovering himself. He had been cycling since five o'clock that morning, insufflating the splendour of Northern Italy with its tumbling hills and gentle light along the river Arno, but he was now in a complete state of confusion, becoming lost three times in ten minutes.

"Mein Gott, somebody is playing dice with these roads."

The road ran between two stands of poplars, their mottled green tips running like fingertips through the ethereal blue canopy of the sky. Olive trees and vineyards dotted the hillsides. Farmers waved to Einstein as he dodged the stones and mantraps of the avenue's potholes. His wheels knocking and jolting the handlebars, he swore he would never ride a bicycle again after this holiday: but he might consider a tricycle. At that point, the front wheel blew its inner tube and Albert rode his bicycle to a standstill.

"Gott!" He swore as he wiped the sweat from his very new moustache.

He was still a young man, with curly, almost frizzy hair and bright eyes; and attracted interest by being uninteresting. His greatest claim to fame back home in the small town Ülm, was being able to drink a large Bavarian beer and throw it up faster than anybody else. His mother often said of him, that it was like raising a German boy, which he was oddly enough. Albert Einstein was as a youth, simply exactly the same as everybody else, with no curiosity at all for physics: at least until this crucial day.

A car came roaring up the road from behind him, in as much as an automobile at the fin de siècle was able to roar, really more of a slow putter-putter and kerchunk. It stopped and ground to a halt on the yellow gravel beside young Einstein. The driver of the car was an adolescent Erich Weiss, lately from Hungary, a member of a travelling carnival and also touring the rural area. He invited Einstein to ride with him, who after looking at his bicycle readily accepted.

Weiss, as he quickly explained to Einstein, was making his way to America to become a great magician and hypnotist, perhaps the greatest the world had ever known. Einstein, naturally grateful for the ride and not wishing to offend, complimented him and wished him all the best for the future.

Unknown to both, however, Erich Weiss was the world's most abysmal magician and hypnotist, and had in later years to change his name to Harry Houdini and became the world's greatest escape artist to elude the countless law suits that were to be brought against him for his extraordinary blunders he made during his carnival years. Nevertheless, this was far in the future and neither of them had any inkling of what was to be.

Young Houdini noticed, by Einstein's clothes, that he was a German and hence a foreigner like himself in Florence. To this, our hero applauded Harry's perspicacity and asked him if he wanted to watch as he, Albert, drank a glass of Bavarian beer and threw it up again, this being Albert's only claim to fame. Houdini, a sensitive Hungarian, declined, but made an offer for Albert to be a part of the carnival for one afternoon, helping Houdini in one of his performances - by being a stooge from the audience.

"Ja, ja," said Albert eagerly, "may I bring my bicycle?"

Latter at the performance, Albert found himself sitting in the front circle of the circus, surrounded by the local contadino, rough peasants in their smocks and work boots. He was jiggling up and down in nervousness at the thought of playing this clever joke on this medieval minded parish, people very much like himself, in fact, and wondered if he should take his bicycle with him into the ring.

Eventually, after the clowns failed to make the peasants laugh, the tumbling acrobats had not fallen over, and the lions had fallen asleep. The young Harry Houdini entered the arena. In those days, his stage name was the Great Salami, which often threw audiences into much confusion, as they could never see the sausages he was purportedly selling. Dressed in flowing white cape and sequined top hat, he seemed very much the part of an itinerant sausage seller and was often stopped in the street because of it. With a grand flourish of his hands, Houdini greeted the audience.

"This afternoon, I will give a demonstration of the stupendous mysterious and arcane arts of Mesmer's hypnotism, the odylic force of animal magnetism passed down to me from the very oculus of the Great Svengali."

Einstein and the audience gibbered in incomprehension.

"It is necessary to ask for a volunteer from the audience."

Instantly the entire audience got up and began shouting exuberantly, waving football flags demanding to be taken, with the exception of Albert, who was still trying to figure out what Houdini meant by the word 'odylic'.

"I can't seem to see anyone," repeated Houdini, looking annoyed at Einstein, who was just beginning to realise it was time to go forward.

Clamouring to be chosen, the crowd poured into the ring and began performing tricks of their own, such as juggling, speeches and pulling at Houdini's cape.

Houdini walked over to Einstein and glared, "Is there nobody here who wishes to join me on stage?"

The crowd started dancing the Tarantella about the circus ring, carrying effigies of the Holy Madonna, the local brass band played trumpet music, and above it all, a small break away group began staging an opera by Verdi in the flying trapeze.

Einstein, mildly put his hand up and was yanked by Houdini onto the sawdust to the consternation of the rest of the audience. Truculently, the simple peasants resumed their seats, darkly muttering about the unfair advantages of German tourists. Albert found himself with limelights pouring their glare at him, no longer sure of what he or why he volunteered in the first place.

"Today, I will demonstrate the extraordinary power of the hypnotism over the human mind. I will give this simple farmer boy, this bucolic ignoramus, this egregious moron, a mind equal to the greatest of geniuses! But first, let's us have the Deus Ex Machina! "

An enormous Catherine wheel was lowered from the canopy to the expectant audience. Upon its finish, obscure symbols were painted, tantric badges embossed its rim and in the centre a large golden star, whose flames radiated along the spokes turned into the mouths of giant serpents.

The audience clapped in appreciation, at about the same time as Einstein almost peed his pants in fear.

"This wheel is the engine of eternity, of life and death, the hidden coil within the Universe! It is the ancient wheel of the mystic sages, of Brahma and Buddha, of Hsi Wang Hu and Yen Lo, of Moloch and Marduk, of Triboluminence and Roget's Unabridged Thesaurus. The Wheel of the Meshuggenah!"

Einstein fainted.

Shortly after, he woke, to find himself strapped onto the presumed Wheel of Meshuggenah, spinning a dozen meters off the ground, as below Houdini enchanted and spelled, supposedly hypnotising young Albert.

"Gott," Einstein strained, as the whirligig span round, "why didn't I become a rabbi?"

"Silence," roared Harry, who was concentrating on building up the tension in the crowd and making strange flourishes with his meretricious cape. He pulled on a rope attached to the rim of the wheel spinning it round. "You are becoming sleepy, your thoughts are draining away, your mind ..."

"Let me down!"

"Your mind is drifting into the clouds!"

"I think I'm going to throw up. Can I have a beer, now?"

Understandably, some of the spectators moved out of the front stalls and out of the circumference of danger, which had began to grow as the heavy wheel started swinging through the air, its massive size discernibly pulling at the set of ropes suspending it from the canopy, dragging guy pegs from their holes.

"You are insensate to the physical world," unyielding, Harry continued, his incantations and spinning, determined to make a spectacle at any cost. "You have entered the quintessence of the logos! Come my child, you are one with the Cosmos."

Thereupon, Houdini slowed the wheel and returned it to the floor of the circus ring.

"Volia! Ecco!" shouted Houdini as he untied Einstein from the Catherine wheel and guided him with difficulty before the audience. Naturally Einstein was still spinning, his legs flying left and right and would have seen the world whooshing past his eyes at sickening speed, if through no intention of Houdini, Einstein had not just been effectively and completely hypnotised: his mind really had been turned by the whole experience. Though, it was to be expected, for as his mother would have put it, 'Turning any corner, would turn young Albert's mind.'

Houdini brought Einstein back to the centre ring and proceeded with the rest of the show.

"The Granda Salsiccia will now demonstrate the power of the hypnotism over the human mind," turning to Einstein and making passes over his head, he whispered the answers to the questions he was going to ask our befuddled hero.

Einstein, of course, had no idea of what was going on, let alone giving a reply to the impossible questions that Houdini was about to submit him to, indeed, a small trail of drool was beginning to edge its way down his chin.

"You have the mind of a great thinker. You can answer any question, as easy as breathing. Knowledge is to you, the stuff of life itself."

The drool made its way to Albert's vest.

"Randomium, Rhenium, Rubidium, Rutherfordium.
Pointillist, Psychoanalyst, Nutritionist, Vivisectionist,
To you all these questions, answers seem hideous
Within your actions, answer as a Physicist!"

It was neither pentametric nor iambic verse, but it had the right feel, Houdini especially like those words ending in 'ium', the way they rolled off his Hungarian tongue.

"Now we shall ask this simple fellow the great mysteries of the Universe!" Houdini played to his audience. "Has anybody a question they wish to ask this simple peasant?"

In the audience were scattered members of the circus troop, also confederates, who were to raise their hands and provide stooge questions, but all were beaten to the point as the mayor, school teacher and star football player of the town jumped into the ring, waving a football banner and bounded over to where Einstein was swaying back and forth.

"Signore. What are the three great laws of Sir Isaac Newton?"

Einstein simply stood there, his blank eyes focused on the distant canopy of the tent above, the drool staining his vest with digestive juices.

"Every action has an equal and opposite action," Houdini hurriedly whispered in Einstein's ear, at the same time wishing he hadn't taken the back route from the hotel.

Inside Einstein's mind, there was a great cavernous hush, the emptiness of eighteen years of Bavarian upbringing echoed ominously, although in a distant part of that enormous space the word beer did kept muttering to itself somewhere in that solitary splendour.

"WHAT ARE THE LAWS OF THE UNIVERSE?" Thundered the mayor, tripping over the pole of his football banner as it got caught in his official red sash.

Finally, after a great quivering of Einstein's German lips came the almost silent reply. "Bier."

Perplexed, the mayor asked the audience. "Bier? What is this bier?"

Houdini tried to hide himself in his beautiful long flowing cape, then with a strained smile explained, "It appears … it appears this particular subject may require a little more treatment."

"Bier?" repeated the mayor.

Contritely, his eyes drilling an escape tunnel into the ground beneath him, Houdini walked the drooling Einstein back to the Catherine wheel, while muttering death threats in his ears. All the while, the uninhabited region of Einstein's brain was finding itself uncomfortably lonely and other parts were wondering if it was time to start throwing up. Then, a rather pathetic iota of consciousness started complaining about being tied again to that unreasonably motile wheel that was suspended up in the middle of nowhere.

"Wharrg!" Einstein yelled out again, as he found himself soaring around the circus tent, with Houdini below screaming out obscenities in Hungarian.

The hawsers and rigging that held up the pavilion began groaning under the sideways motion of the heavy wheel. The tent poles now visibly moving in about sockets above the ring.

The crowd began making earnest glances at the exits behind them.

"YOU ARE GETTING SLEEPY!" yelled Houdini, dragging vehemently on the rope attached to the wheel, "YOU ARE ENTERING A TRANCE!"

The fabric of the big top began billowing, waves of cloth shimmering under waves originating from the tips of the poles, the ropes attached to the pegs in the ground, one by one, were popping out of the ground, the whole edifice of the tent swaying in rhythm to the motion of the lurching Catherine wheel. As Houdini, his face red with exertion dragged intractably on the cabbalistic engine's tackle.

The audience - half out of their seats or half out the door - were muttering Ave Marias, Paternosters, and extracts from Dante's Inferno, deciding it was time to leave.

"YOU ARE ASLEEP!"

Finally, and as surely as the earth swings around the Sun, as naturally as homemade organic ice cream, or as impossible as the women's auxiliary rugby team winning the national finals - the Catherine wheel collided with one, then the other, the poles supporting the Big Top, snapping them in half. The tent fell with a wonderfully great 'whoffling' sound, around Houdini ears, and the afternoon's entertainment fluttered to a halt.

Einstein, still tied to the spinning wheel, bounced off the

dancing elephant and shot out the curtain door, following the screaming peasants into the Florentine Campagna. Literally cycling along the road, and racing towards the town, he awoke to find an epicycloidal world rushing past his nose, with a vague hint of impeding doom as terrorised people ran in all directions. He wondered with lurching uncertainty, if this experience was anything like Ben Hur felt in the coliseum, and more importantly, "Where are the lions?"

A chase began, as the wheel passed through the town, clusters of old men and women took up arms against this demon apparition, which tore through the piazza and market place before heading out along the main road towards the river. Lines of freshly dyed sheets were mowed down by the onslaught of Einstein's monster, the town dogs fled yelping down sides streets in mad panic, flower beds were ploughed under, and a six hundred year old renaissance statute by Giotto was atomised by the wheel's iron rim. The wheel eventually hurdled the bank of the Arno River, disappeared into the water for a moment before reappearing downstream with Einstein still tied to its side and floated away from the cataclysm it had left in it wake.

A week later, Einstein, crazed with spinning around and around, was dragged out of the river, at the town of Pisa and immediately arrested for wheel smuggling, which at the time was a capital offence in rural Italy: but he was latter released on account of impressive ability to regurgitate beer.

And this is the true story of how Albert Einstein found his scientific genius and muse.

Einstein and Siggy

Herr Einstein visits Herr Doctor Sigmund Freud in Vienna for a consultation, after being lost for four year - he loses his way even further.

Albert Einstein stood outside 19 Bergasse Strasse, Vienna. The house was hidden down a long winding cobbled street, between over-leaning houses of white panels and richly carved balconies. High above the road, in gabled windows of leadlight glass, old women sat leaning on flower boxes filled with marigold flowers. They stopped their discussions between the balconies and stared at the odd sight of Einstein with his trusty bicycle, upon which he still carried the hiking gear he had used in Italy. Tied to his bike were a blanket, a pair of boots, an old felt hat, a statue of the leaning tower of Pisa and unused dirty post cards. He paused and took account of the house's entrance. The portal showed a marked difference from the others in the street, with strange hieroglyphs and mandalas carved deep into its surface. A large brass plate, done in old Germanic script, proudly announced the consultation rooms of one:

Herr Doctor Sigmund Freud
Vienna Surgeon, Hypnotist
Psychoanalyst
And
Wart Remover

Einstein frowned and looking sideways, considered taking lunch in the nearest beer keller, so as to work up enough courage to enter this acclaimed doctor's practise, but also he considered the possibility of practising his own claim to fame, that of beer regurgitation, a sport which he found required constant training to stay in competition.

The young, and hitherto unknown, Einstein had made his way to Vienna, following Houdini's bungled attempt to hypnotise him for the purposes of a carnival show. Strangely, given all that had gone wrong, the hypnotism had worked, against all the expectations of Houdini. So well in fact, that Einstein's brain was now twisted sufficiently to make it impossible for him to function normally. Giving him bizarre dreams, filling his mind of things of which he had not the faintest comprehension, but most importantly him famous beer spewing had been profoundly affected.

Now, upon the portal of the slightly famous Sigmund Freud, Einstein was having second thoughts and began to turn away, when the door flew open and he found himself—and his bicycle—dragged inside.

"Come in, come in, my boy," Freud said, motioning for Einstein to step into his study, "here take a mummified scarab beetle."

Sigmund Freud—or 'Siggy the Loony' as his parents had insisted on calling him—was a small jittery man, whose eyes, hidden behind tiny glasses, would dart round the room as you spoke with him, refusing to meet your gaze, he would constantly point out objet d'arts he had carefully scattered throughout his office to draw attention away from himself. A meticulous dresser, in black suit and cravat, he often gave the impression of a raven hopping about his room quickly giving clients bits of pottery from Africa and Asia to examine, while he hopped behind them and picked up more pieces to exchange as soon as they had finished with the last. This would continue on for some minutes, until the clients began to feel dizzy and would ask to lie down for a minute, at which point, Siggy would offer them the 'couch' and begin his consultation.

"Pardon?" Albert replied nonplussed, "Are you Herr Doctor Freud?"

"How about this lovely druidic sacrificial dagger, Eight Century to a day, thought to originate in Saxony?"

Then he dodged behind Albert to grab a fragment of Grecian pottery, circa Herodotus.

"Err," Albert span round to keep up with him, "I'm interested in discussing ..."

"Did you know Omphalos is originally a Greek word signifying 'navel' and refers to the centre of the world, as situated in Delphi in ancient Greece?" shoving a Grecian urn in Albert's hands then bouncing across the room to fetch an icon.

"Do you mind if I sit down?" Albert tested, "It's just I've had a bit of a turn lately?"

Freud froze in the centre of the carpet, a vague smile half smile forming on his lips, turned slowly, then he rapidly bounced back across the room, scooping up a pre-Columbian Mayan sculpture of a dog to deposit it in Albert's growing collection. He gestured intensely at the couch by the wall, took up a seat directly behind Albert and began smoking a cigar.

Einstein, lying on the couch with his arms full of bric-a-brac, rolled his eyes trying to come to grips with Freud's odd behaviour. Twisting his head round to speak, he found Freud jumping to his other shoulder and thrusting up a small Russian icon into the flowering collection, before shooting back out of sight. Einstein thought, perhaps, all these trinkets might be part of the treatment, but felt this hard to accept, as he hadn't mentioned what his dilemma was yet, but more importantly he thought, as he stared at the sacrificial knife, he didn't qualify as a druid.

"Herr Doctor Freud," he hesitated, thinking another curio would appear psychodynamically in his arms, "Every night, I have been having these strange unearthly dreams that make me wake in a pool of sweat, as if I am trapped in a sweatbox or terribly sick with influenza."

"Good," said Freud and took a tug on his cigar.

"Good?"

"Yes, very good. No, please don't get up. Here have this. Yes, now tell all of us about your dreams."

"Us?"

"Sorry, what was I thinking of, tell me about your dreams. So just relax now, and let your mind do the talking."

More perplexed than ever, Einstein waited for an understanding to arrive as he stared at all the objet d'culture lying in his arms, he began to bite his nails and wriggle his feet, till, out of nervous irritation he began to describe the events of his summer holiday in Italy and his mental collision with Houdini, the demon hypnotist.

"So you see," he finished, "ever since they pulled me out of the river Arno, I've been having these strange headaches during the day and terrible nightmares at night. I am at my wits end."

There was a silence, as Einstein waited Freud to reply. There was none, and the expectant silence filled the room with a disconcerting void. After several minutes, Einstein felt tempted to twist his head round, just in time to discover Freud tiptoeing back into the room with a cup of coffee and a plate of Vienna cream buns.

"Like a bun?" He giggled thrusting the plate into Albert's hand and immediately jumping back into his seat, out of view, whereupon he asked, "These dreams, do you enjoy them?"

"Well no, not as such," Einstein said and began munching on a cream bun," I mean they have a sort of eerie emptiness to them, as if I'm floating round in space, like what happened when I tied up on the cartwheel, and they possess this absolute conviction that there is some ultimate and godlike meaning to it all. It's not as if there was something to enjoy, it's just strange, really."

"Sorry, I meant are they filled with lots of sex?" Unseen by the reclining Einstein a perverted leer twisted Freud's face. "Do you have to clean the sheets in the morning?"

"Err … no, no, not at all, there is just this space floating around me, or me floating around this space."

"Ah, so, yes," said Freud disappointed, "ah well, it is so obvious to me young man, that you have gone completely insane."

"Oh, I hadn't realised."

"Yes, of course, but this is not a hopeless case. We will have you changing your sheets in no time. Ja, ja, no time at all. But first we must come to grips with the unconscious forces that repress your latent, hysterical sexuality. Tell me, have you ever been seduced by a vulture?"

"A vulture?" Einstein started forwarded on the couch, only to be handed another Russian icon.

"Yes, you see," Freud, chortled in his element. "I remember Leonardo da Vinci was seduced by a vulture as an infant. It is a common occurrence in Italy. Especially along the river Arno, which was the same region you travelling through. And remember the story of Leda and the swan; there are so many examples of this avian lechery taking place in history. It would not surprise me in the least," Freud gesticulated with a finger in his coffee, "if all these flying dreams are simply because you were seduced by a vulture. That is my considered opinion."

"That's incredible, I never would have imagined. Yes, I can see it now. Herr Doctor Freud this is so astute of you to discover this so quickly. But, tell me what about the hypnotism?"

"Hmm… no, we are trying cocaine. It was in the bun I just gave you. How do you feel? Ecstatic, sense of wild exuberance—

erection maybe?"

"Cocaine?" Einstein felt a surge of panic overwhelm his bafflement.

"Yes, clinical cocaine, I have found it very effective, but of course terribly addictive. Tell me about your dreams?"

"But Doctor Freud, perhaps ... whoa!" The surge of cocaine hit his bloodstream and was now doing wonderful things to his brain. "The colours ... the lights ... the sounds ... the erection."

"Ah, gut, the erection is so unmistakably important in dealing with sexual repression in males."

"But I am not sexually repressed. I'm quite normal ... ohhh! "

"Rubbish," interrupted distinguished Freud, "everybody is sexually repressed. It's a normal state of mind. We are meant to be sexually repressed, that is not the important concern, but we still need to recognise why."

"Why?"

"Why? But, of course, you do not understand these things, I have been studying them for years and still I have not the faintest idea either, so do not worry 'why'. It is more important for you to know 'when'."

As the drugged bun took greater hold of his conscious and subconscious facilities a broad grin was seen to evaporate Albert's concern with the present, and he began giggling in ebullient pleasure.

"W ... when ...he he he ...what ... ha-hahaaaaa ...ohh? What that me?"

Freud shifted his chair closer behind the couch and began taking notes.

"Ah, so, now we begin understand the neurosis of your Oedipus complex, ... or is it an Electra? Ach, never mind. Now, you must solve the Sphinx's riddle of your mind."

Einstein stood up and began to wander round the room. His head commencing to sway from side to side, and his pupils dilating to black opals: a marvellous vacant look grin beamed from his blissful face.

"I feel so light and free," he sang, falling onto the wall and collapsing as a heap of clothes on the Persian carpet, burbling, "It's like flying," as he started crawling his way along the wainscot before colliding with a roll-top writing desk. Crawling beneath this, he hid his head at its back and came to the stunning and irreproachable conclusion that he was in reality—a cockroach.

"A cockroach, you say?" Freud repeated from his recliner, "A cockroach, now that is very interesting. Ja, very interesting. A cockroach is, you know, a small nocturnal beetle-like insect infesting kitchens, of the order Blattaria. But what make this especially interesting is that I have another patient in Prague—a complete loony —who also thinks he is a cockroach. Herr Kafka. Only he doesn't need the cocaine to release this bit of his mind, on the contrary, he really believes he is a cockroach. Oh dear, perhaps you have regressed too far?"

Einstein began chewing on the varnish beneath the desk.

"We will try something else—mein lieber. I want you to tell me the first thing that comes into your mind when I tell you a word. Okay?"

Einstein clicked his heels together, which for some deranged reason, he believed were his antennae.

"Light?" prompted Freud, taking the first idea that came to mind.

"Bumble bee!" came a loud, though muffled reply.

Freud made a note of this.

"Dark?"

"Bluebottle!"

"Time?"

"Dung beetle!"

"Space?"

"Stinkbug!"

"Gott," Freud wiped in face in consternation, "—what a schmuck!"

"Gott!"

"Gott?"

"Schmuck!"

"Schmuck?"

"Gott!"

"Ahhhhh!" Freud murmured thoughtfully.

"-Choo!" Einstein replied congruously.

"Gesundheit."

"Danke."

Freud opted for a respectful silence at this funeral of a human mind.

Presently Einstein shuffled out from under the desk. Flecks of wood varnish clustered about his lips and chin. His ovoid eyes swam across the room, searching—perhaps—for a lintel he could flap his insect wings and fly across to. Freud drew his legs up on his chair and hid his face behind the notebook, all the time scribing tiny notes in the margin. Scuttling forward, Einstein charged across the room into the hall, making tiny 'eeping' noises as he did so. His hands and feet turned outwards in imitation of a cockroach, and like the wings of the insect, his coat tails flapped wildly behind him. Then past his waiting bicycle he turned right and somersaulted down the stairs into the kitchen cellar.

Erupting into the kitchen, he threw himself on Freud's wife, Martha, and tried dissolving her by spitting digestive juices over her face. She screamed and knocked him off with a Polish sausage, belting him several times over the head and endeavouring to throw a burlap sack over this mad rustling creature. A Dalmatian dog began barking wildly, running in circles round both of them, trying to bite both Martha and Einstein, the circles became smaller until in a fit it bit its own tail and fell over. Einstein still floundering under the impression he was a insect, tried flying through Martha into the larder beyond, only to trip over the dog and collapse on top of the Martha, who, with as an elegant riposte stepped out from underneath him and boxed him to the floor with the sausage.

"Sigmund! Sigmund! Is your patient staying for dinner?" She cried out, and then walloped him again for good measure.

In a wave of lucidity, Einstein rolled up to his feet, staggered back up the stairs, knocking down Freud who was trying to hand him a stone age statute of fecundity, grabbed his bicycle and burst out of the front door and took off down the twisting street trying to find a place that sold good Austrian beer.

Darwin the Vivisectionist

Evolution is how Nature cooks – Charles Darwin

Charles Darwin was seasick, he had been seasick from the first dawn, and he had been so relentlessly seasick for a week now that he was beginning to wonder if his intestines were being attacked by a sea cucumber. This was surprisingly close to the truth, as the ship's cook had been poisoning the entire crew ever since he had come aboard. The cook's approach to cuisine was his revenge on the British navy for having press-ganged him into service from the finest restaurant in Belgravia and carted off to serve slop to drunken sailors aboard the HMS Beagle. Most of the meals consisted of whatever odd animals he could drag from the sea, ranging from albatrosses, vampire squids, and most provocatively ambergris - which for lack of a better title was listed on the menu as vomitus whaleus.

Darwin stared out the porthole. Plymouth sat there as steady as a rock and he wondered if they would ever set to sea. The captain had insisted that before their voyage the officers and gentlemen spent time with each other, it had only taken one meal for them to develop an instinctive dislike of each other and of the dining room. The captain assured them that seasickness was common even among seasoned sailors, and then threw up a curry made up of octopus eyeballs, leafy sea-dragons and malaria tablets.

"Gentlemen, a toast!" captain FitzRoy lifted a glass, "to a successful voyage, and to sweethearts and wives - may they never meet!"

"Could we just have a slice of dry toast?" Darwin groaned.

"What's wrong with the meal?"

"Well, either it's still alive or it has come back from the grave,"

Darwin pushed the plate away.

"Nonsense," FitzRoy looked annoyed, "the cook is the finest we could pressgang from the restaurants of Belgravia. None finer."

"None finer than what?" Darwin watched amazed as the shell of some antediluvian crustacean emerged of its own volition like the Birth of Venus from beneath the soup.

"Well, you know, none finer." FitzRoy regarded Darwin circumspectly, "Charles, we will soon be at sea on a voyage that may last up to five years, a little discomfort is to be expected in the cuisine."

"I would most certainly agree captain," Darwin felt green, "except we're still in Plymouth harbor and have yet to set to sea. We could at least have our meals in the tavern, while the cook decides whether to poison us or kill us with scurvy."

"Nonsense," FitzRoy thumped the table with his fist, "the Royal Naval has a tradition to uphold, and the food is part of that tradition, no matter how foul it may be."

"The other thing," Darwin looked under the table at his ankles, "is it really necessary for all of us to be shackled to the floor?"

"That also is part of the Royal Naval tradition," FitzRoy waved his hands about to dismiss any arguments, "all new sailors are to be shackled while in port, this prevents deserters from well, deserting you know."

"But FitzRoy," Darwin complained, "we're not Jack Tars, we're officers and ordinary gentlemen."

"Enough of that!" FitzRoy stood to attention, "we will sing Heart of Oak, the anthem of the Royal Navy."

The rest shambled to their feet, dragging their chains and FitzRoy lead them on with first verse:

"Come, cheer up, my lads, 'tis to glory we steer,
To add something more to this wonderful year;

To honor we call you, as freemen not slaves,
For who are so free as the sons of the waves?"

Not knowing the words to this, all Darwin could do was rattle his chains and look uncomfortable.

The next morning they assembled on the deck and Captain Robert FitzRoy read them the riot act.

"It has come to my attention," he looked down sternly at Darwin from the poop deck, "that there are rumors of a mutiny on board my ship. Given that we have been sitting in Plymouth Harbor for only a week, and have several years of sailing to get through, I have decided to set an example to you all."

Darwin felt the earth give way beneath him, and then he realized it was just the swell of the sea. He watched FitzRoy with the sense of alarm.

"I have decided to show that your captain is just as capable of enduring the rigors of the sea as any of you are, and to this end," he waved at Bosun to come forward, "I am going to have myself keelhauled. Bosun!"

"Sir? What?" The Bosun looked at the captain in alarm. "Isn't that a little extreme?"

"Nonsense!" FitzRoy used the sort of stare Nelson gave with his right eye, that is to say – blindly. "Attach the chains to my limbs, and have the men haul me under the ship."

"Beggin' your pardon sir," the Bosun bobbed his head and looked to the junior officers for support, "but we can't be doing that sir, not unless the captain has committed some offence again'st King's regulations and what like."

"And King's regulations state, the captain is the final arbiter in all matters." The captain took off his hat and jacket and handed them to the Bosun, "now carry out your orders, or it will be you, and not I, that goes under the ship."

This was too much for the Bosun and he quickly tied the chains to captain FitzRoy's arms and legs, and had the ordinary seamen toss him overboard. Several minutes later he was dragged back on board, none the worse for the object lesson.

"There see," FitzRoy spoke directly to Darwin as he wiped the seawater from his face, "the rigors of our journey are to be endured, not complained or whined about mister Darwin."

"Yes captain," said Darwin, in no little awe of this demonstration, then he pointed at FitzRoy chest, "are those lampreys or eels?"

'Waah!" FitzRoy jumped in the air and started flailing at the creatures, which had attached to his torso. "Get them off! Get them off!"

That evening the gentlemen and officers reassembled in the officer's mess and were presented with a meal that could best be described as 'inimical'. Darwin pushed uncertainly at his plate and watched animated green bubbles make their way to the surface of the gruel only to decide death was the better alternative, before collapsing once more into the mire.

"What is this?" he asked as the mouth of a scientifically impossible fish open in the gruel and took a gulp of air before also dying. "Is it cooked?"

"Mister Darwin," captain FitzRoy put down his spoon and gave Darwin an unrelenting look, "you are again setting a bad example to my junior officers. Please refrain from doing so."

"But this food, if it could be called food," Darwin poked once more at the soup, only to have it attack his spoon, "is it still alive? I mean seriously, I'd swear those are teeth."

"It is present on the table," FitzRoy was not giving an inch, "it is in a bowl, it is steaming and it has nourishment. Therefore, as captain of this ship, I am declaring this food. Gentlemen – the King."

They all raised their glasses, as Darwin was heard to mutter: "Like large teeth, the short you see on a terrier. Is this terrier?"

"Mister Darwin," FitzRoy thundered, "The King!"

"Ah, yes, the King," Darwin joined them in the toast, "Although, I certainly wouldn't expect King William to eat something like this."

"Mister Darwin!" FitzRoy slammed his hands on the table and rose to his feet, "I certainly hope, you do not continue in this manner during the voyage of the Beagle, or it will be a very long voyage indeed!"

"With food like this," Darwin looked suspiciously at his food and continued to mutter, "I'd be surprised if the voyage lasted longer than a week."

"Darwin!" FitzRoy threw a key on the table and pointed at the door, "Leave!"

"Ah," Darwin grinned a little, took the key, unlocked his shackles, and then crept out the door. This was his whole plan all the time, only it meant he could slip a shore and grab a proper meal from the tavern. In the back of his mind, he hoped the Beagle might sail without him on the morning-tide.

The next morning they assembled on the deck and Captain Robert FitzRoy read them the riot act – again.

"It has come to my attention," he once more stared angrily at Darwin from the poop deck, "that there are rumors of dissatisfaction with the ship's menu. This menu was carefully selected by the cook to sustain us during our arduous struggles in Southern Ocean."

The cook, who was about to strangle a cat in the gunnels, met this with a look of surprise.

"Given" FitzRoy continued, "that we have been sitting in Plymouth Harbor for only a week, and have several years of sailing to get through, I have decided to set an example to you all. Again!"

Once more Darwin wondered what lay in store for him.

"I have decided to show that your captain is just as capable of enduring the rigors of the sea as any of you are, and to this end," he waved at Bosun to come forward, "I am going to eat a cat of nine-tails. Bosun!"

All the Bosun could muster was, "Wot?"

"Bring me a cat of nine tails, a pot of boiling water and a fork," FitzRoy did what he thought was a heroic impersonation of Nelson. "I'm going to eat a cat of nine tails, if I can do that, then any meal no matter how bad will be deemed edible."

"Beggin' your pardon sir," the Bosun bobbed his head and looked worried, "but as we only have the one cat-o-nine, it would seem to be a bit of a loss when it came time to serving King's Regulations to the crew and all. Can't we be using something else instead?"

"Have you any suggestions?"

"Well how about the ship's cat," he pointed at the cook, "going to be served up anyway."

"What?" FitzRoy stared in amazement at the cook with cat in hand, "that's what happened to Matilda?"

"What of it?" the cook threw the cat on the deck, "You wanted fresh meat! I'll give you fresh meat!"

The cat raced up the ship's rigging and hissed at them from the yardarm.

"You've been feeding us cats?" captain FitzRoy was amazed.

"Cats, dogs, seagulls, rats, hairy crabs, glass frogs, eels eyeballs, toad legs, vole lungs, and haddock," the cook spat on the deck, "but you only got haddock because we were out of sturgeon brains and blobfish."

FitzRoy coughed for a moment, then leaned over the rail and vomited into the sea. A flock of seagulls dove down to investigate then flew off in terror. FitzRoy recovered and waved his arm at the cook.

"You are dismissed from service!" he croaked and returned to wave watching.

"Finally," the cook whooped and threw his hat in the air, before racing down the gangplank and disappearing.

That evening, a rather green captain FitzRoy joined his officers in the tavern. There was a minute's silence as they watched him expectantly. He raised a glass of port and stood at the head of the table.

"Fox hunting and old port, ships at sea," he said simply and then looked at Darwin. "Tomorrow we will be having a new cook, trained in the French style of cuisine. The King!"

"The King!" they chorused.

Darwin raised his glass and grinned to himself. It has taken a whole week to convince the cook to serve up offal and call it food, but it had been worth it and the voyage would now have a decent cook, and if it didn't, he had an even more cunning plan.

"The King!"

Newton and the Case of the Crown Jewels

For every action there is a reaction.
And for every hero, there is a villain. - Newton in his cups

"The street's clear," a short swarthy fellow lifted cowl from his head and pointed down the street.

The street was indeed clear, as a storm had swept it clear from the people and refuse that normally cluttered. The year was seventeen hundred and one; it was England in one of its quieter moments between the wars and without the plagues. The present monarch wasn't a homicidal maniac; he didn't have pretensions of godhood and spent most of his time over in Europe trying to evict French squatters from Holland. Most of England was filled with commonsensical folk who had only two purposes in mind, becoming better shopkeepers, or getting drunk and whoring; most of England, however, did not include a tall distinguished looking Cambridge professor who went by the name of Isaac Newton. Newton, in the rump of his years, had taken up crime solving with such fervor that his friends worried if he was suffering from a permanent brainstorm.

He appeared from the shadows and loomed in the night.

"Then let us proceed," he said to his companion and valet Hedby, "and quietly, for we must not be discovered, or all will be lost. I tell you truly, that dark deeds are afoot and only careful footwork will see us right."

Hedby looked at his shoes, then at his masters and then back to his own feet.

"But I polished them, I did sir," he protested.

"Quiet man, not our shoes, I mean the deeds of evil."

"Are we buying a property?"

"Silence you blithering fool, England is at stake."

"I didn't know you had wager on the footy, sir?"

"One more comment like that, and you'll be back working in the morgue. Now as you said, the street is clear. Let us this night set right a greater wrong than this land has ever seen."

They made their way along the ways of the dismal avenue. This was an age without street lights or police, it took a stout heart and a stouter cudgel to walk the night streets for fear of a dagger in the back, or blackjack to the back of the head. Many a poor soul had disappeared in the murk of the Great Wen of London, and Newton was determined not be one of them.

He had come here on this dark night on account of his position as Warden of the Royal Mint. Charles Montagu the Chancellor of the Exchequer who had informed him of a calamity that threatened the whole nation had visited him the three days before.

"Somebody's nicked the crown jewels!" Montagu fell at Newton's feet.

"Get up man," Newton drew back alarmed. "We don't do that sort of thing at Cambridge, at Oxford I'm sure they do, but here there is no buggery at all!"

"Newton for god's sake, you must help me!" Montagu clawed at Newton's gown and almost pulled him to the ground. Hedby, who was coming in the door carrying a jar of porter, neatly averted his eyes to the ceiling and abruptly pivoted out the door.

"Come back Hedby! It's not what you think!" Newton jumped back to the fireplace and picked up a pair of fire tongs to defend himself. "And Lord Montagu, kindly refrain from physical contact or I shall be forced to use these!"

Eventually Hedby was able to sit Charles Montagu on a divan and quieted him down with a jug of porter. Montagu raved for a few more moments until the ale settled him and he sat looking aghast at his host.

"Now then," said Newton, "it would be best if you proceeded along a more rational approach in this manner and calmly explained yourself with at least a measure of decorum that befits a minister of the state."

"Oh thank sir," said Hedby as he started to sit on a chair by the door, "don't mind if ..."

"Not you, you blithering idiot!" yelled Newton chasing Hedby out of the room with the tongs and slamming the door on him. "Blasted servant! Now then where was I? Are yes, Montagu, what on earth is the matter with you?"

"It's the jewels Newton, the blasted crown jewels, they're gone missing. That's what the blasted matter is. I say this ale is rather good."

"Missing, you're sure of this, you haven't simply mislaid them?" Newton stood by the fireplace and began packing a pipe with tobacco.

"My dear fellow, they are a little hard to misplace, given collectively they weigh half a tonne. Made of gold, silver and god's knows what else. In addition, the fact they are kept under lock and key at the Tower of London does make it a little difficult for them to suddenly just disappear. No, I fear they have been stolen."

"Well, if they are missing, it would make sense they are stolen."

Newton stood before the fire and with the tongs pulled out a glowing ember to light his pipe.

"Yes, quite."

The great scientist puffed at his pipe and gazed upon his stricken friend.

"And what endeavours have you taken to search for, or

recover these missing jewels? Have you enquired at the pawnbrokers or gem sellers? That would be the logical place to look for them."

"The king, refuses to believe they have been stolen, and has ordered a thorough search be made for them, in his strong belief they are still within the tower. He refuses to believe his own jewels, even though they are missing from a high tower, surrounded by guards and under lock and key have been illegally taken."

"Yet you believe they are stolen."

"There can be no other interpretation, they are gone."

"And have not the guards seen anything, no locked damaged, no ropes or scaling equipment discovered?"

"Nothing," Montagu mopped his face with a handkerchief. "No marks on the doors, the jewel chests or even on the guards."

"And have you tortured the guards?"

"I threatened them with a pay cut, but they maintained they had seen nothing or heard a sound. I believe them."

"Yes, a pay reduction would put a sense of terror into any man. Now then, who else has access to the tower?"

"Myself, king William and Doris the cleaning woman."

"And what have you gleaned from the charwoman?"

"Only that the guards were drunk on the night, and playing Dwyle Flonking."

Newton bit down so hard on his pipe stem he broke it and cut his lip.

"Dwyle Flankers, now that is almost a crime in itself." He threw the remains of pipe in the fireplace and began packing another. "And the woman, has she come into new wealth recently? Bought a castle?"

"How could she buy a castle? Oh, I see. No, and we searched her, all she has is a pig."

"Pig or piglet?"

"A large porker she had been fattening up."

"Then it is not her, for to remove the jewels would take considerable time and effort, and I doubt very much if she could remove almost half a ton of jewellery with just a bucket. No, and it also would require foresight, it is very unlikely she would continue fattening a pig once she had hers hands on even the smallest of gems."

"Also, she doesn't clean the jewellery room, just the privy out the back and the kitchen."

"Yes, that too," Newton frowned. "I wish you'd said that before."

"Sorry, Newton."

"And the keys to the room, who has a copy?"

"Myself, the head guard, his majesty and Doris the charwoman."

"Wait," Newton twirled and pointed at Montague, "why on earth has the woman got keys to the room if she doesn't clean there?"

"That's where she keeps the pig."

"Confound it man, does she or doesn't she have access to the jewels?"

"Well no, only myself the guards and the king are allowed in there."

"So why in heavens name is she keeping a pig in there?"

"Well, since the crown jewels are missing, the guards said she could till they were found again."

Newton rolled his eyes, then relit the new pipe and walked over to the window to stare at the courtyard of the Great Court of Trinity. He could see the gardener cutting the grass from around the sundial; the man hovered low to the ground, his hat almost reaching his nose and his scissors clicking in the Cambridge peace. On the windowsill was a small mirror that reflected the sun onto the ceiling of the room, where another sundial was painted. Newton was obsessed with sundials.

"It is a singular fact." Newton blew a smoke ring, "that the king refuses to believe the crown jewels are stolen even though they are missing. Most odd, but then again he is a foreigner."

"Newton!" Montagu cried out, "He is also our king and your majesty!"

"True," Newton turned and faced his guest, "and possibly a thief."

"Consider your words carefully Newton." Montague said as he put down his cup of porter. "Or I will arrest you on treason."

"Montagu listen to me," Newton walked over, grabbed the arm rests of Montagu chair and stood over him. "The facts are the crown is missing, since presumably you are not responsible, since the guards face a death penalty and Doris the cleaner only has a pig to worry her. Then the only possible person left is the king, unless of course you can show someone else had access to the room."

Montague stood up and grabbed Newton by the shoulder.

"Sir Isaac Newton, I arrest you for high treason. Guard!"

"What?"

Montague's accompanying guard came in and put a double arm lock on Newton.

"Damn your eyes!" he yelled as he struggled, "I was being hypothetical!"

Newton was imprisoned in the Tower of London with the pig. The pig was fed twice a day by Doris the cleaner. Newton wasn't.

The head guard was a former yeoman who had served with four kings in France, Belgium, Holland and Wapping. He was often known to comment that some of his fiercest battles had been in Wapping. On the hour, and on every hour, he or his deputy would bang on Newton's cell door for the changing of the guard, after two days of this Newton protested that there was no need as the crown jewels were missing, - to which the guard replied he was guarding the pig and to ask if Newton had happened to see the jewels anywhere in the room.

It was only on the third day that a visitor came to see Newton.

"Hedby!"

"Yes, milord." Hedby pulled off his cowl, and nodded. "I was wondering whether you'd be coming home for the weekend, or if I could have some time off?"

"Damn your eyes, man!" Newton railed at him, "I'm in the tower for high treason! What do you think?"

"Well, sir," said Hedby as he scratched at his fleas, "I don't rightly know. Is that a good thing or a bad thing?"

"It's about as bad as it can get!"

"Well, in that case, milord, you won't be minding if I did in fact have a week off, or maybe a couple of years?"

"Damn your eyes, I have to get out of here."

"Well, milord," he jingled a purse, "I took the opportunity of bringing your house-college treasury with me. I was thinking you could bribe your way out."

Newton grabbed him by the jacket.

"No wait, I have a better plan."

"Will it involve me being set on fire like last time," Hedby said with resignation. "Or drowned in the Thames like the time before that?"

"Better still, this time you're going to be disguised as a pig."

"Oh milord," said Hedby depressingly, "that would mean I'm going up in the world."

"Now take off all your clothes and snort like bacon."

"Milord?"

That night Newton was able to escape from the tower by dressing in Hedby's rags, carrying a bucket and tying a string around Hedby's neck as he trotted behind. Hedby to his disgust was covered in muck.

"'Ere now, Doris," said the sergeant of the yeoman as they approached the Traitor's gate. "What you leaving with that there pig for? Ready for market is he? Odd time of night to be headed for markets, it is."

"Oh, yes, fat and ready is he," Newton muttered in a shrill voice, his head wrapped in Hedby's cowl. "Got to get there afore break of day, I do."

"Funny looking pig that."

"All that rolling about in muck," rambled Newton desperately, "you know, pig in the muck kind of a pig."

"Aye," the guard rubbed his chin, "that would be it."

Hedby oinked and cheekily bit the guard's leg.

"Get orf!" yelled the guard and chased them out the gate.

They fled down the riverbank and the across London bridge towards the South side of the Great Wen - a place of blackguards, whores and bishops.

"My time in the room was not wasted, Hedby," said Newton as they redressed themselves.

"What you made friends with the pig then?" said Hedby as he pulled his rags back on.

"Silence fool," Newton clipped him round the head, "I discovered that the room had not been broken into, and from interrogating Doris and the guards I am convinced that none of them is responsible for the missing jewels. Thus I am convinced the king must be responsible, but for what reason I do not know, nor do I know how, and this we must investigate."

"Or lord Montague," put in Hedby pulling his cowl back over

his head.

"Don't be a fool, Montague came to me, and he is completely above suspicion."

"Yes, but milord, he also put you in the clink and by doing that he got rid of the only man in England who could solve this case."

"Damn your eyes!" Newton stamped his feet, "I knew that, I did! I did! - Damn, no I didn't!"

"So, milord thinks maybe Montague may be responsible after all?"

"Shut up," Newton frowned, "I need to think about this possibility."

"Right you are milord. I'll just stand here, guarding I will. Won't disturb you a bit. Mind you there is an inn just down the way, and now that I come to think of it, I'm feeling a bit peckish, and now that I think of it, this is the whoring part of town and I do like a bit of whoring and I was thinking that ..."

"Shut up man!"

"Yes, milord but I can't help but thinking, that my lordship might be better able to solve this problem if he had something to eat and all. And while we're about..."

"Damn your eyes!" Newton shouted in exasperation. "Very well, where is this blasted inn?"

"Thought you'd see it the logical of it, milord. This way, now I do happen," he continued on in his endless prattle leading his master.

They wandered down the dark avenues when suddenly a storm broke upon them, forcing them to flee the rest of the way to the inn.

The inn was one of the most disreputable establishments in one of the most dangerous parts of the underside of London. It would be difficult to find a place even upon the continent where as much crime and skulduggery took place as in back rooms of the George and Dragon. Famed as a brothel where no vice, no perversion was too low or obscene, it was well favoured amongst the landed gentry, writers and murderers as a nice place for a bit of tit and an ale.

"What ya have luv?" the barmaid, if she called be called a maid, swabbed down the bench top and picked some meat from out her teeth.

Newton had one weakness, if it might be called a weakness or simply an infirmity of the brain - he was terrified of Women and the nature of Women in general. They were his bête noire, his Moriarty, his Grendel, and considering women consisted of half the population of the world, it made his task of crime solving and living in general no end of impossible.

"I'll... I'll... I'll..." Newton stammered and looked at the woman fearfully.

"Well?!" she demanded slapping the cloth back over her shoulder.

Hedby came to the rescue. Well acquainted as he was with his master's foible and used it to his every advantage whenever he had the opportunity arose.

"My master would like a porter and a spotted dick, and I'd be havin' the same."

"Wot's wrong with 'im?"

"Gout."

"Gout?! Never seen gout do that' to a fellow afore."

Newton just stared wall-eyed at the woman's bosom and cringed back in his chair, wishing he were back in Cambridge where the opposite number in gender was banned by order of parliament, however, she left in a moment and he let out a sigh of relief.

"Damn your eyes, Hedby, why did we have to come to a brothel?" he struck his manservant with his hat. "And what on earth is a spotted dick?"

"It's delicacy milord, made from pudding and flies." Hedby grinned. "Not the sort of high meal they serve up at the Great Hall back in Cambridge milord, but just what we need right now."

"Porridge and flies? Nobody eats flies, surely you mean something else."

"No milord, oh wait, maybe its fleas I'm thinking of."

"Flies or fleas, I'll have none of it." Newton wafted his hands. "Just make sure that dreadful creature doesn't return."

"What dreadful creature?" She asked as she put down the platter of drinks of spotted dick.

"I… I… I…," he almost fainted.

"There was a pig here afore," Hedby smirked. "He's worried it may return."

"Pig," she pondered, "must be from the second floor. Here's ya meal. That'll be sixpence."

Hedby pried open his master purse and paid her, all the while Newton seemed to shrink further and further under the table.

"Er'e now? Wot you looking fur?" she demanded, "O'im respectable I am. Keep ya eyes off me mittens."

Newton breathed easy, as she stomped away probably the only barmaid on the South bank with a sense of propriety. He looked at the spotted dick that Hedby was flensing and poked his own with a spoon.

"Those are not flies," he picked up one and looked at it closely, "Hedby, this is a currant or what is better known in Latin as a Ribes rubrum, in fact it is a red currant, and by the looks of it a French red currant, of which, it is floating in a bed of what appears to be egg custard with rhubarb."

Hedby sighed, seeing the second bowl of custard slip away, it was the gaze of a lemming as it contemplates the last bit of iceberg it is standing on, as it dissolves slowly to nothing.

"Yes, milord, I expect you'll not be eating it then?" he said without too much enthusiasm.

"On the contrary, I expect I will be!" Newton took a mouthful, ruminated on it for a moment. "Great heavens, this is delicious. Why on earth did you think it had fleas in it?"

"Oh, just a rumour I heard."

"Gads! What splendour!" Newton smiled happily, "tell why we are never served such wonders back in Cambridge?"

Hedby looked worried.

"Well milord, due to certain fiscal arrangements that have been protracted by previous merchants within the bounds of the Cambridge markets and their environs there is a shortage of ... um ... recipes in the University colleges, so to speak."

"Recipes? Recipes?" Newton held up his spoon and for a moment considered his manservant, he was about to speak when the barmaid returned and he looked up in alarm.

"You want anything else?" she barked.

"I ... I ... I ..." Newton gawped with his spoonful of custard and completely forgot what he was about to speak.

"Ah, no thanks lass," Hedby let out another sigh. If his master ever discovered where the money from the college's kitchen was going, he knew he would soon after be hanging from the Newgate pillory. "Milord, perhaps you should give some thought as to the matter of the missing crown jewels?"

"Yes," Newton snapped, "I know that! I do! I do!"

"Perhaps we should investigate the manor of Lord Montague?"

"Are you mad? That would be tantamount to breaking the law,

actually that would be breaking the law. I've half a mind to arrest you, but I fear that if I did I'd be rearrested in the process."

"Milord, I said investigate, not break in."

"Oh, well damn your eyes, I knew that!"

"Yes milord. Will you be having that ale as well, milord?"

"Be quiet, I have to think."

Soon after, they found themselves in the street, discussing 'setting right a greater wrong than this land has ever seen.'

Hedby pointed up at the house opposite.

"I believe this is the manor of Lord Montague, milord."

"Nonsense Hedby," Newton snapped. "I know my way though London, we are nowhere that part of town. This is Southwark or my name is Reginald Clovis, I'd stake my reputation on it!"

Hedby merely pointed at the sign hanging out the front of Montague house.

"Ah." Newton's face had the expression a whole coastline of pre-Columbian civilization might have had after it had just been wiped out by a tidal wave. "He must have moved recently then. Yes, that would be it."

"Yes, I expect your right milord." Hedby remarked sarcastically. "This being the ancestral home and all of Lord Montague, I expect he wanted to move home again."

"Yes, quite right," Newton tapped Hedby on the shoulder, "I see my intellect is rubbing off on you. Good show."

The building was a white washed thatch affair with most of the ornamentation being confined to a small shoe scrapper on the stoop. The residence, along with many of its brethren in the street, maintained an Anglo-Saxon no-nonsense approach to housing that harked back to the days when people had names like Ethelbald, Ethelbert, and Elthred the Unready. To say it was uninteresting would be doing a disservice to the person who had discovered the word 'uninteresting'.

"Milord what shall we do now?" Hedby asked.

"Since we can not enter this house, without Montague discovering us, we should question his neighbours. And by doing so discover if anything strange has happened. This maybe the only way."

They knocked on the house opposite and a man wearing a kilt and an angry expression opened the door.

"What the bleedin' ell do you want?"

"We are doing a survey for ..." He never finished as the door was slammed in his face.

"Milord," Hedby asked, "Should we not accost the fellow for such behaviour?"

"I hold it a rule never to be rude to a man wearing a kilt," said Newton as he walked to the next door, "either he has an extended family in the form of a clan, or is stark raving mad. Either way, you would not wish his enmity to last beyond a wind change."

A more congenial man who invited them in and gave them supper, under the misapprehension they were fellow Quakers, owned the next house.

"Come in gentlemen I have been expecting you," he lead them into his parlour, which was a grim puritan affair consisting of well made straight back chairs and a puritan humor to go with it. "Would you gentlemen like some cake? Most of the other Friends of the Society have left, but I am always happy to engage in fellowship with my brethren."

He offered them a slice of hard baked porridge, and Newton banged his on the table to determine its density.

"I thank you sir," once again he banged his biscuit of porridge to see if it left a dint in the tabletop, "but we are here on the investigation of a crime."

"A crime?" the eyebrows of the Quaker shot up like a Methodist suddenly discovering he was at a Latin mass. "Sir, there has been no crime committed within these walls, unless it be illegal to worship our lord with prayer."

Hedby also banged his biscuit of porridge on the table and muttered about illegal cooking.

"Fear not good, sir," Newton reassured him and giving his hard tack to Hedby, "our inquiries deal with the house opposite, the manor of Lord Montague to be precise."

"Well, sir, I wouldn't be knowing anything about that. Well there's nothing that your sirs would think important. And truth be known, sirs, I would never give testimony against another."

"Hmm, I think you may know something but choose not to, please feel free to speak your mind," said Newton, who was a man committed to the principle of free speech, but had never been to Russia where it brought a death sentence.

Hedby took the biscuits and dropped them over the side of the table to a small terrier that wisely ignored them.

"Well, sirs, since you have insisted, there was a strange thing that did happen five nights ago, it did. I was out front of my house, here, and was proclaiming the word of God to all the poor wretches who live in the darkness away from…"

"Yes," Newton interrupted not wishing to listen to a religious lecture, "to the point man."

"Well, sir, I saw this great carriage approach the manor of his lordship Montague. Now I don't right know whose carriage it was, but I do rightly know that it was not a simple fellow. Rather one who is high in the world and far from the word of our savior who brings the light…"

"And who got down from the carriage?"

"Well sir, I saw tall dark haired man who spoke with a foreigner accent, he did. It was not the sound of a good English man who followed the light of our saviour ..."

"Yes, yes!" Newton interrupted once more in exasperation. "Could you please explain the matter without constant recourse to our Lord and Saviour, I'm quite sure he has enough on his hands to deal with, without listening to your incessant prayers."

The man brought his hands up to his face and thought about this for a moment.

"Why yes, sire, I do expect your right, I had never considered that God might expect us to deal with matters by ourselves without constant appeals for help."

"Quite."

"On the other hand, what sort of all power and all seeing god would not have time for each and every soul, unless of course... wait, no, unless he wasn't .. but that would mean... zounds! That would mean I have wasted the last twenty years of my life in fruitless prayer... Nooooo!"

"Er," Newton held up his hand, "that's not exactly what I meant!"

"This is too much! I can not bear this idea!"

"Steady on, old man. I was just wanted you..."

"Great merciful Jesus, all the crumpet I could have had!"

At this the Quaker tore open his shirt and ran out the front door screaming.

"Nicely done, milord," quipped Hedby.

Newton slapped his forehead.

"There is nothing more pointless than having a conversation with someone who believes in the principle of eternal damnation. Not only are they damned boring, but damned if you can learn anything interesting from them."

"Well at least we know there was a carriage with a foreigner."

"Yes Hedby, there is that."

"Very well, let see what else we can learn from this street."

The road consisted of three parts, that which was bog, that which was road and the bit they were standing in, which could reasonably be called a morass, the sort of liquid which would entertain someone herd sheep in the west of the Shetland Isles. As they stood there, they saw a great carriage approach and stopped opposite them. To their surprise a man in a black lacquered mask stepped out, he looked at them possibly with equal surprise and then waded his way across the street to the manor of Montague.

"Milord?" said Hedby to his master.

"Yes, I saw. He is wearing a mask."

"No milord," Hedby grabbed Newton's arm, "I meant I fear we are sinking into the mud."

"What?"

"Mud milord."

"Damn your eyes! Hedby!" They found they were stuck fast in the mud and were sinking quickly; soon they would be up to their knees. "Why didn't you say sooner?"

"Well milord, far be it from be to assume milord wouldn't notice sooner."

"Quick, lie down in the mud, it will spread the weight of your body. Quickly man."

Hedby lay down in the mire with a countenance of resignation that one normally perceives on the face of the condemned.

"Now stay there."

"Yes milord."

Newton lunged on top of him and began dragging himself through the mire to what could conveniently be called a doorsill.

"I'll just stay here, milord."

"Yes, now be quiet."

"Thank you milord, could you make sure Timmy gets a chance to go to school," he mourned as he sank deeper into the murk.

From across the street the carriage driver looked at them unsure if they were a pair of drunks, a pantomime act or escaped lunatics. He looked up at the sky and perceiving there was no full moon, he determined they must be drunks.

"Er mates, hang on I'll throw you a rope and drag you out," he called out cheerily.

"It's quite alright, my good man, almost there," Newton panted as he dragged himself forward.

"Throw the bleedin' rope mate! Before they stick a bleedin' tombstone over me!" yelled Hedby.

Soon after they were free, covered in much from head to foot, stinking of sewage and nevertheless alive. Newton took this opportunity in hand and used it to question the driver about his passenger.

"Well, mate," he winked, "I can't tell you who is he. But he's ain't a regular fare and all, that's the god's honest truth, but I will tell you he always wears that mask. And he's always wearing all that perfume and all. Let's see what you make of that."

"Hear that Hedby," Newton pounced on this remark. "That passenger has a habit of wearing a lot of perfume."

"No milord, I was too busy getting the mud out of me ears."

"This is very interesting. Yes, I would go so far as to say, I almost know what is going on here."

"It's a road mate, that's what going on," the cabby sniggered.

"Come Hedby, it is time I cleared my name."

"Try cleaning your jacket first matey!" The cabby fell over himself laughing.

They entered the house without waiting to be let in, the sound of viola and flutes echoed through the stately mansion.

"Newton! I arrested you!" Montague yelled.

"Yes, and you arrested the wrong man!" He walked across the carpet leaving a trail of muddy footprints behind him. "Here is the fellow, to whom you should have placed in the tower!"

He pulled off the black eye mask and revealed a stout man about fifty with beaded eyes and a large almost condor like nose staring angrily at him.

"King William the third of England, King of Ireland, the Scots and Stadtholder of the United Netherlands! This Montague is the man."

"Yes Newton," said Montague, "I'm quite aware who the king is, as so are the rest of my guests. Now kindly explain yourself, while I call for the night watch to come and rearrest you."

Newton stared about the room and saw to his surprise a half dozen more masked people sitting in the candlelight, all looking in amazement at him. It was as if in the middle of the third act of a Shakespearean play the lead actor had started singing an opera in German.

"Your majesty!" Newton bowed to the floor, "where are the crown jewels?"

The face of King William was like the expression an alpine village would have had after an avalanche had fallen on it from the side of the Matterhorn. He first looked at Montague and then his guest and then helplessly tossed his hand on the table.

"It's true, I know where they are."

"Aha!" Newton looked triumphantly at Montague, who in return stared with amazement.

"But how did you know? How did you discover this?" The King asked perplexed.

Newton paraded about the carpet, bit of mud falling off him and decorating the linen.

"It was elementary." Newton strutted. "To have removed that quantity of gold and jewellery without anyone noticing required a vehicle which has wide rims on its wheels. And there is one such vehicle waiting outside at this very moment. Secondly, its user was in the habit of wearing perfume, which as you know is a continental conceit. The king has spent most of his life on the continent. It did not require too great a leap of intuition to discern the king might be responsible."

"But the guards would have known the king was there," protested Montague.

"Not at all, for on the night the crown jewels went missing from the tower, those guards were given a whole barrel of ale. I know this for a fact, for during my own incarceration at the tower I was able to interrogate both the guards and the cleaner. And in was in that same barrel the crown jewels were hidden and rolled out of the Tower. This I deduced from the depth of the ruts made by the barrel leading into the stronghold and back out."

"Well done, master Newton, well done," the King clapped and in accord the other guests did as well.

"But why, why on earth should the king steal his own jewels?"

Newton smiled and picked up the mask he had lifted from the king.

"To pay for the wars against the French, he has pawned the jewels, with the intention of returning them at some point in the future. This is why he insisted that the jewels were not stolen, only missing. The king has not broken any law of the land; he has simply borrowed his own property with the intention of returning it. Is this not so, your majesty?"

"Quite correct, master Newton," the king stood and bowed

expressing his admiration of Newton's abilities. "Now you are to be locked up again at the Tower, until such time as I am able to do so. I cannot have the world know of my financial dealings. Guards!"

"What? No, I say…"

Newton was dragged off and returned to the high dungeon where he remained for six weeks before the jewels were returned. In that time, he became very familiar with a pig.

Einstein and Adolf

Beer and Politics make for strange bedfellows - Adolf Schicklgrüber

Einstein charged into a beer hall, grabbed the first foaming tankard off a table and drowned his Adam's apple, sighing, he wiped his hand across his mouth and gazed blissfully about the hall. It was festooned with Austrian and German flags; men in lederhosen and brockwürst slapped each other cheerfully on the behinds, grapevines hung from the rafters and barmaids carried armfuls of beer steins to the guests.

Einstein sighed contentedly, Freud may not have cured him but at least there was beer.

A nascent Aryan political group gazed appreciatively back at the appearance of this healthy, young, beer, swilling German, and marching to the beat of a different drummer boy, was a twelve years old, pasty faced Adolf Schicklgrüber, who came over and clapped Einstein on the shoulder.

"It is gut. Ja!" said Adolf.

"Ja!" looking down at this strange youth with a stein of beer in his hand. Albert replied enthusiastically, never having heard of the Aryans. "Beer is the best!"

"Would you like to join the Aryan Beer Drinker's Party?" A beaming smile broke on the face of the child who would later change his name to Hitler.

"Do they drink beer?"

Einstein, having spent the morning in Sigmund Freud's Clinic force fed cocaine and having his neuroses and higher brain functions successively: tweaked, jumped on and thrown out the cognitive window. Was now suffering from too much thought, and desperately required inebriation to obliterate the first buds of higher intelligence.

"Ja," beamed a jovial Adolf, who was yet to grow a toothbrush moustache, "and many other social activities. Are you German?"

"Ja," Einstein happily replied, quaffing back another stein of beer, "I was born in Ülm, near Switzerland. I have just finished school."

"Me too. And, are you Aryan descent?" inquired Adolf, a look of uncertainty clouding his eyes.

"Ja," Albert thumped back, and woozy from the effects of Freud's cocaine had the impression 'Aryan descent' meant throwing up: which he promptly did.

"Gut!" thumped Adolf again, wiping Einstein's mess off his shirtfront, "we only want pure Aryan Germans in our new Aryan Beer Drinker's Party. Isn't that right - men?"

A group of squat, short, low browed, black haired, brown shirted teenage boys had clustered around Albert and Adolf. The air was steamy from warm camaraderie, hands filled with large jugs of beer and hot sausage. Their dark brown eyes sometimes purple and bruised, sparkling in warm contentment, often above broken noses and missing teeth. Giving a friendly hurrah, and thumping each other on the backs, they hustled their newfound comrade out to the beergarten, at the back of the inn.

The tall buildings of Vienna rose majestically above a tableau of trestle tables and noisy revellers in the cool open air of the outside beergarten. A perfume of fragile white jasmine flowers hanging on old oak beams, disguised the smell of stale sweat, stale beer, stale cigars, stale meerschaum pipes and the fresh vomit that was the only thing fresh to the garden.

Pretty, young Germans maidens -actually pretty, young Serbian emigrant workers with blonde wigs- carried large platters of beer, wine and Swabian noodles to happy, hungry Austrians, who rocked shoulder to shoulder, merrily singing 'Edelweiss' and 'A

Spoonful of Sugar Makes the Beer Go Down!' in overly loud voices. In contrast to which, by the door, a three piece band of tuba, clarinet and piano-accordion were trying to play Vienna waltzes, in between being thrown out of the beergarten by irate skin heads, whenever the band refused to play the noisy bits from Wagner's 'Ride of the Valkyries'.

"I want to hear Wagner!" yelled a red faced brown-shirt at the band, "Why won't you play Wagner?" and poured another glass of beer over the unfortunate tuba player.

"Ah yes," said Adolf happily to Einstein, "We must have our Kultur. Where would the German people be without our great art and culture?"

The other guests of the garden included large, stout men in lederhosen and alpine hats grouped in a corner. An undernourished scullion who ran along the tables collecting used plates, and was constantly tripped up by young Adolf and his merry men. As well as, a physics professor and his students from the university crammed into a corner, drawing equations on the tabletop, by dipping their fingers into their beers and running their foaming fingers across the wood.

The room was a bedlam of vociferous bawdy groups, tramping up and down on stained paving stones, which crisscrossed the floor with little runlet's of beer and wine, trickling beneath them carrying bits of meat and blood. A seat collapsed beneath the weight of too many drinkers, sending them flying into the isles, knocking down one of the beer maidens, who threw her platter of Swabian noodles high into the air covering the beleaguered band with tiny winding-sheets of floppy spaghetti. Einstein found himself dragged wholesale, to the corner of the room occupied by the physics professor and his students, the students engrossed in their equations.

The professor in question was Ernst Mach, a gentle, seventy-three years old self-effacing man, and renown for his extraordinary intellect, his high culture and his sense of superlative moral virtue. He would often bring his students down to this part of the town, to be part of the rich social life it had to offer, drink lots of beer, but most importantly because he was madly in love with one of the beer maidens. The professor was engrossed in the weighty movements of a certain serving girl named: Mileva.

Mach, world celebrated for his humanitarian invention of the ballistic missile, had fallen in love with her, from the very instant he first laid eyes on her and she had punched him for putting his hand up her skirt. Laid out flat in the swill and trampled food, the professor stared up at this tough, uncompromising woman with her glaring eyes, sleeves pulled up to the elbows, wild red hair hidden beneath the obligatory blonde wig, and swearing at him in violent Serbian - in a way that should have turned milk into acid. Instead of hiding under a table, he had smiled disarmingly, his blue eyes glistening from pain and asked for another beer. She took a step back, disconcerted with his pathetic remonstration, picked up a jug of beer and threw it at him, knocking him once more to the ground.

Unfortunately for Mileva, it was of no avail, for, from that moment on, Professor Ernst Mach lived only for her, and her extraordinary ability to knock a man out without even blinking.

A tall splendid woman, whose ancestors had been Cossacks from the steppes of Russia in the seventh century, who after storming into Southern Europe on their way to the Italian Riviera, had ended up conquering and becoming completely debauched in Constantinople, finally chose lives as itinerant psychotic mercenaries and dentists in the Northern Balkans. The women of these people have remained so fierce, that their name in translation is taken to mean 'walk through walls'. And this -at least to Herr Ernst-gave Mileva a transcendent, ethereal quality- seriously like the vocational end of a swinging battering ram.

He was so engrossed in Mileva slamming down steins of beer on the other tables, that he failed to notice the sudden arrival of the Herr's Schicklgrüber, Einstein and the youthful alliance of Austrian storm-troopers. Which, after a moments consultation, conducted in heavy breathing and momentary grunts, had pitched the university students out on their ears and ensconced themselves at the table. Mach was jammed into the corner next to Einstein, followed by an exuberant Hitler and a group of neophyte Nazis-who immediately began playing such drinking competitions as 'Spot the Table', and 'Great Farts of the Weimar'. Old Mach tried waving frantically to get Mileva's attention, but found all her concentration was wrapped up in grabbing another drunk by the head, twisting it down between his legs, and propelling vigorously him towards the door.

"We must make all of Greater Germany into one nation," yelled the teenager Hitler, his huge nose thrusting into Einstein's face, "that includes Holland, Austria, Denmark, Poland, Russia and most of South America, including that funny bit, you know … Australia."

Einstein, quite drunk, was staring down at the sets of equations crudely written on the table in drying beer foam.

"Beer," he said profoundly, and nodded his massive head sagely.

"We must insure the purity of the German Volk," unperturbed the child continued thumping the table with his small hand, "there must be no Jews or foreigners in Germany."

"Ja," Albert's body replied indulgently, "There must be no Jews in Germany. Even I who am Jewish, think there must be no Jews in Germany, in fact there must be nobody in Germany … only beer." The body had attempted what it thought would be an apposite quip and drank another tankard.

An extraordinary silence began at the table and spread across the room, like the silent film of a shock wave from a nuclear explosion the instant before the cataclysm erupts. Every head slowly turned and focused on this ill-starred remark. With the exceptions of Mach and Mileva, one absorbed in the other and the other absorbed in not being absorbed by the other. A feeling akin to social eczema began to permeate the hitherto merry atmosphere, as people began edging away from the table.

The old professor Mach, became round eyed when he sensed the mounting storm and tried ducking under the table, as it was heaved upward then tossed high across the room by the bevy of fascist students. For an instant, he thought of hiding behind the fool who had made the remark, but opted for hiding under his seat.

Einstein stared heavily at the place where a moment before his beer had been. The thought - 'huh?' went through his mind for a millisecond, before he surmised he must have drunken it. Another part of that cavernous interior, prompted him with - 'more beer!' He looked around and found to his surprise, the entire room watching him with disturbingly intent.

"Beer?" he piped quizzically, as he was picked up six stalwart fascists and hurled about the courtyard, very much like a happy, young harp seal playing a friendly game of catch with a pod of killer whales.

After his fourth flight across the enclosure, his head began to tingle and feel decidedly blank, small points of light coursed about his pupils, and a loud ringing hum felt all along his spine. For an instant he thought he was going to explode in a burst of fulminating energy: then he threw up and felt much better.

Einstein, suffering the combined effects of eight steins of beer; Freud's dose of cocaine and inexplicable psychoanalysis; the recurrent bouts of delirium of Houdini's hypnotism and now playing the flying Zucchini Brothers with Martin Bormann - suddenly found his mind jump utterly and completely out of his body and begin strolling round the room, much to consternation of his flying cadaver.

And now, for the first time in its utterly vacuous existence, Einstein's brain found it could think clearly, and not only clearly, it was thinking brilliantly. Everyday observations about the constitution of the world became shattering realisations on the fabric of the universe, space and time. Watching its body get heaved once more through the beergarten, made it realise the extraordinary importance of relative motion and absolute frames of reference. It found shattered itself on the floor, the table with the equations left behind by Mach, which might explain the creation of the universe and the possibility of a Big Bang.

Mileva, after watching the frolics of the young Nazis with Einstein's body: felt moved. Moved - not out of any inkling of tenderness or sympathy for this soaring rag doll, rather because she felt she was missing out of the fun. Cross-eyed and raven hair with a blonde wig, she strode through the brawling chaos, knocking the young ebullient Nazis helter-skelter with her fists and feet, grabbing them occasionally by the collar and shaking them like puppies till their senses feel apart. She grabbed Albert in mid-flight by the feet and carried him upside down to the door, past the band and threw him out into the street. Dusting off her hands and spitting profusely in the gutter.

Einstein attempted to stand for an instant then with a moaning retch, adroitly heaved the last of his beer in a glorious parabolic curve onto Mileva's blouse and fell back down into the cobbled street.

She picked him up again and shook him like so many Swabian noodles and in a way only explicable to the Slavic mind decided to fall madly in love in with him. Laughing in amusement, she dropped him on his feet once more in the drain and like a tempest stormed back inside.

Albert's mind after wandering errantly through the beer house, found itself outside Albert's body and contemplated with studied intensity: the possibility of staying outside. Yet before it could choose wisely, it was abruptly sucked back inside the maelstrom of his body: screaming abuse at the fates for throwing it into such an intellectual abyss. He stood up like broken glass, a thousand thoughts passing through his mind, a million emotions quivering with incomprehension, the ramifications of truth and existence poured via his soul to the very basis of his being, and for the first time in his unbelievably asinine existence he began to understand.

Mach came running out, white hair flaring in his passage, threw his hands out, to violently grab Einstein by the neck, knocking him once more to the ground.

"She likes you," he screamed, "how could you?"

"She chucked me out?" Einstein replied, throwing off Mach's pitiable attempt to throttle him.

"Yes-but she's never done that to me! Why you?"

"Moustache?" Albert stroked his lip.

"She doesn't have one - I must know." An earnest Mach ranted wildly, "Tell me, what is your aftershave? Have you read Zola? Are you still a virgin?" He turned and gibbered with unrequited love. "She is extraordinary, a Vesuvius and Venus, a Daphne and Damocles, poetry in action, she fills my world with a radiance that illuminates the darkest night, with a light of purity and nobility which is as soft and warm as the golden sun dissolving the hollow shell of the night, as she rises serenely above the morning mists. Her smile is as tranquil and soft, as the parting of purest white clouds suffused with the pink glow of a spring dawn. All of heaven is outdone, the stars are reborn burning afresh, the four winds turn and begin anew bending toward her gaze, and the earth pauses in its orbit and revolves about her."

She threw another body out the door, and it landed on top of them. Einstein and Mach disentangled themselves out from under twelve-year-old Schicklgrüber's madly flailing limbs.

"A grace unhampered by ties of gravity," continued Mach unperturbed, "her arms and neck are as liquid thoughts bending to change the world, hair glistening flax cascading upon classical shoulders the most perfect of artists will never capture. She is princess and goddess, yet more real, more alive, more true than the blood that flows in my veins and the air that I breath..."

The tuba player fell with a heavy thud on top of his euphonium, followed by other members of the band, the rest of the youthful Nazis and the stunned mayor of Vienna.

"All ... I asked for: was a plate of sauerkraut," he stuttered in amazement, brushing down his mayoral uniform, and shaking his head.

The group of men stood around and discussed how to retake the inn - whether a blitzkrieg attack, subterfuge or waiting for her work break would be most likely to win the day. Mach ran amongst them, gesticulating feverishly, trying to change the subject and the means, his white hair becoming more and more awry, one moment hanging on the lapels of the mayor and the next trying to box young Schicklgrüber's ears to make him feel reason.

Einstein stood on the edge, knowing he had passed some Rubicon in his life, though uncertain as to what it was, and mostly puzzled as to the fact he was able to think.

Just as they were about to storm the door by sheer weight of numbers, the Valkyrie appeared once more, holding above her head an enormous hundred litre barrel of wine, which after swearing in vivid Serbian and giving a withering look of apocalyptic vengeance - hurled it down, scattering them like dwarfs before a titan. The wine barrel exploded throwing up geysers of red, quenching the heat of troops and finally removing any question of the battle's victor. Then she bent down, and throwing Einstein over her shoulder.

"Come we will go," she said.

"Where?" he moaned.

"Transylvania," she smiled and walked down the road, whistling an old Cossack tune, "and mama."

"Mama?"

"Mama and marriage!"

"But I just wanted a beer!"

Three weeks later in a Slavic whirlwind marriage he was married.

Einstein and School

Before Einstein was a genius, he was a child.

The Kindergärten was old, older than the ground it was made on; this was a contradiction, which the school managed to ignore. Corbels and flying buttresses soared awkwardly to nowhere above the city street and rose back down in complete defiance of local gravity and city ordnances. Pedestrians scurried past, taking with them a feeling of unease, which they could never identify and would not shake for the rest of their day. Pigeons would alight for a moment and then fly off in a whirl of feathers and fly non-stop to Morocco. In the year of our lord 1683, Mustafa the Turk had invaded Europe with 150,000 blood thirsty Mussulman and charged right onto this city, and upon taking one look at this building had charged right out again, vowing never to return. There was something odd about it.

A distracted boy and his ovoid mother stood in front of the edifice, holding a scrap of paper between their hands, they looked up curiously at its refusal to collapse under the weight of an uncanny architecture.

"Mother," asked the distracted boy, "why is that boy in chains?"

"You may be assured Frau Einstein," barked the headmistress Fröbel, when the Einsteins were let into her office. "That we give the strictest adherence to discipline, logic and beating. Classes begin at six a.m., precisely, with an educational swim through the river and five-kilometre jog with full pack and rifle. This is followed at seven a.m., precisely, by the breaking of fast with one potato raw, a bowl of porridge which is frozen - weather permitting -, forty push ups and a stein of beer, those boys failing to consume their beer must immediately recommence the five kilometre jog with full pack and rifle, followed by another stein of beer. Classes commence at eight am, precisely, with various subjects including Latin, mathematics, and readings from Clausewitz, Kant, Nietzsche, Frederick Barbarossa and our cherished Chancellor Bismarck. This is followed by luncheon break with cabbage soup, hard bread and beer, during this break the students are expected to recite Goethe's Ode to the Fatherland followed by one hundred push-ups. The afternoon is filled with boyish amusements such as fencing, horse-riding, Greco-roman wrestling and bayonet practise. Afternoon story telling from the Brothers Grimm, Jacob and Wilhelm, is followed by nap-time at three sharp - students failing to sleep will be beaten.

"All students are expected to swear to serve the Fatherland on the Russian Front, those students who choose not to, are subject to full court martial and forced to stand in the corner for the remainder of the term."

Frau Einstein and young Albert stood in complete trepidation and hung their heads, hoping to avoid the ominous stare of the übermistress. Of Brobdingnagian proportions, she soared above them, seeming only to lack a spear of Odin and a chariot loaded with heroes slain in battle to take back to Valhalla singing joyfully of their missing body parts.

Her office was testimony to Prussian order and precision, one could almost hear the mice making lederhosen behind the walls, where ornate world clocks of many countries were hung and all identically set to Berlin time. The furniture was pre-Reformation in tortured leather, with manacles and accoutrements, to give sitter a constant reminder of what the chair went through for their comfort. Behind the mistress, a set of paintings in the imperial style, of previous headmistress' serried the wall, - their immobile faces so severe the artists had all retired to convalescence homes soon after the paintings –intense points of focus sprang from their eyes of these portraits, all looking back to a time when men were still übermen and everyone else emigrated, and forward to a time when the gods would fight their final battle of the Götterdämmerung only with a view to using howitzers instead of hammers.

"Here, at the Friedrich Fröbel Military Kindergärten, we strongly believe the growth of the child, using the enlightened teaching methods of my father Friedrich Wilhelm August Fröbel, to strengthen the student's own abilities, through play, beating and mindless discipline. We study the Three F's…"

"Err, you mean surely…the Three R's…"

"No! I do not! I mean the Three F's! Friedrich Nietzsche, Friedrich the Great, and our Glorious Fatherland!"

She glared at them. "Any questions?"

Two blank stares.

"No?" A smile broke like an avalanche on the glacial features, "good, so now we will proceed to the Darwinian Fitness Test."

The headmistress poured down the corridors with Albert and his Mother in tow; students and teachers alike scattered as bathers before a killer whale.

"The importance of the Darwinian Theory of Evolution can not be denied in the evaluation of a student's fitness." Proclaimed headmistress Fröbel, decanting around a corner and into a Frankensteinian laboratory, "Ecce Homo! Test the child and behold the man!"

A cacophonic orchestra of machinery and electrical brushes wired and buzzed, around a seemingly dimensionless device of extremely well polished steel, wire and leather. In the midst of it all, a small gauge electrical railway complete with sedan chair and dental fittings hummed unnervingly.

"This," the head mistress tapped the sedan chair with her

riding crop, "is the zenith of German technical expertise, German artistic endeavour and German dental work that money can buy from the Hamburg Shipyards."

All Frau Einstein could muster was, "Ach!"

"Mein Gott!" Erupted from Albert's mouth as he was lifted up, strapped into the sedan chair, and a leather cap was loosely tied to his brachycephalous cranium.

"And now we test for the Üntergattung, as we recapitulate the stages of Evolution and Genesis." The headmistress declared as she threw an electrical switch, which arced menacingly, and Albert shot forward along the toy railway. "First you will note your Kinder has entered the primeval ocean, birth place of all Life and the German State."

Albert plunged into a long oily tank of water, leaving a trail of bubbles and muted screams as he disappeared beneath the surface. "Here, as you may see Frau Einstein, is the primary test for amphibian nature, note your son's striving to experience the world using his proto-limbs."

The chair and Albert affixed, broke the surface like a resplendent u-boat after a successful convoy sinking.

"This leads us to the test for the navel."

"Navel?" quizzed Frau Einstein.

"Yes, our esteemed researchers at the University of Berlin have discovered that all mammals, do in fact, possess a navel. Hence, the primary test for the essence of mammal Nature - is in fact ... a navel."

"A navel? Really?"

The engine of evolution ran straight up the wall, across the ceiling and onto a universal shunt.

"Yes, the Navel, in the original Greek is called, Umbilicus, which of course, is directly related to the Omphalos which is "the navel" of the Earth…"

The chair of horror dropped from the ceiling accompanied with a wail from Albert, into a large bathysphere, whose lid promptly clanged shut, and the sphere began bucking and gyrating on the spot.

"Here," continued out headmistress Fröbel, "we have combined the nature of the Earth and a representation of the womb into one, note the detailed work on the metalwork."

"Eh, yes… Is he alright?" Frau Einstein looked in horror.

"But of course!" Parried the headmistress, "The device is the product of the finest German Craftsmanship. What could go wrong?"

"An explosion perhaps?"

"Never! Made in Hamburg! There see on the plaque… there above the jars of sulphuric acid for the batteries?"

"Ah, yes, of course."

From within the fiendish globe, broke wild screams of fear and help.

"Mama!"

"Excellent! Notice he's yearning to connect with his mammalian ascendancy. He has passed the test. Any animal that calls for it's mother, must be of mammalian stock."

The capsule broke open and the chair erupted out, spewing the hapless aquanaut once more onto the train of doom. It raced about the room at high speed, a giddy young Einstein screaming his young Bavarian lungs to their full.

"And now we test for Branchiation."

"Branchiation?" question Einstein's mama, who at the same time was a little proud that her young German boy was of 'mammalian stock'.

"The mechanism of Evolution which was first so clearly enunciated by our foremost German scholar Ernst Haeckel, that proto-humans have retained the use of our limbs for the method of climbing, to wit, we dangle from our arms as we climb. Now, note that at the end of track are those monkey bars, which your son, ... oh dear."

Albert, his arms flailing about sedan chair, was swept at high speed along the tracks, his leather cap firmly locked over his eyes making it all but impossible for him to unstrap himself from the engine deus ex machina and leap out to snatch the monkey bars, and instead plummeted in a violent heap at the end of the track onto a pile old mattresses, which he still managed to miss and rather flew straight onto the wall.

"Ach," tutted headmistress Fröbel as she unstrapped him, "two out of three, is good enough for this little one. Come along, I will show you to the boy's dormitory."

"Mama!" The intrepid Einstein wailed.

"It's okay, dear," his mother comforted, "it's for Science."

The boys of the dormitory crowded around the new arrival, who having calmed down sufficiently from the morning's ordeal looked with curiosity at his new quarters.

"My name is Karl Mauch, I am the head boy here. I am most pleased to make your acquaintance." The head boy shook Albert's hand, bowed correctly and clicked his heels.

"My name is Uwe Johnson, I am the second boy here..." began another, and so on through the rank of borders. "And you, new boy - who are you?"

"Oh gosh! Why I'm Albert ... no really, I say, I like your buttons, this is my first day, Mama brought me and we met the mistress, but I do like flowers, and sometimes when it's raining we go to canal and look at the ducks who can swim which is why they have feathers, and now my bicycle is being repaired by the nice policeman and I can get it back after I have beed De..poor..ted... Do any of you have bicycles?"

This unfiltered stream of consciousness was simply too great for the Dormitory boys to swim through and the inquiry was met with a wintry glower of displeasure. The entire group of borders retired to a crook of the room to thrash out a common campaign to deal with the trespasser.

Meanwhile, the young Albert was making himself home by was circumnavigating the dormitory, it was a immeasurable space, with a vaulted ceiling that disappeared in shadows and whispers, twenty beds arranged meticulously in symmetric rows along the length of the room, linen so sternly made up that sides of beef could have been thrown down and cut up on them with little effort, each hard cot was surmounted by exquisitely carved calligraphic aphorisms exalting the boys to maintain hygiene by constant brushing and polishing to dedicate their lives to the Kaiser; stained-glass windows portrayed scenes from The Ring of the Nibelung done to a modern theme with Siegfried carrying a Mauser rifle instead of a long sword, and beside each bed was a tiny library of world classics, including "Boy's Own Teutonic Knights", "The Rise and Rise of Prussia", and "Frederick the Great – I did it my way!".

The boarders reached an agreement and marched in single file down the room to confront their new housemate.

"We have decided you will be permitted to use the facilities." Karl proclaimed folding his arms, "however, speaking out of order is forbidden, the riding of bicycles is prohibited, and ..."

"My bicycle is red and silver, and has a bell which my father made out of a shoe and sometimes Mama lets me ride in the park which has a sea with boats and flowers, and sometimes when it's raining we go to canal and look at the ducks who can swim which is why they have feathers and sometimes when the postman comes by I race him with my bicycle until he hits me with a rock…"

Albert trailed off into silence as the backs of the boarders retreated to consider implementing the Schlieffen Plan, which lacking mechanized infantry they realised was improbable; words like 'annihilation', 'vengeance' and 'Götterdämmerung' featured heavily in their forum as they sagaciously and individually mooted appropriate methods of retribution on the interloper, involving the toilet block, the adjacent canal and the idea that met the greatest approval, of feeding him to the school's vicious guard dogs.

Albert left to himself wandered out to the school yard.

It was a cold day and the birds were absent, though that was mainly due to the Kindergarten's forbidding architecture. From across the quadrangle, a monocled Cyclops appraised him, the eye belonged to a drummer boy who marched at Austerlitz and lost an ear, who saw Napoleon at Jena as the other eye gained an eye patch, charged at Waterloo and lost a knee, fought on both sides in the year of revolutions of 1848 and lost a shoulder, a kidney and a whole army division, he fought the Danish in 1864 and lost his hair as well as the province of northern Schleswig, and in the Franco-Prussian War he was all but killed - but refused permission to do so by his commanding officer and instead captured Paris. The school rifle instructor had seen so many wars that his body had taken up a completely new topology and was the wonder of the local university's medical department, there seemed nothing this Übermarionette could not survive, outwit or simply shoot dead at a range of three hundred metres, that is … until this day.

"You there, boy, come here."

The talented child looked in alarm about him, there were many shapes about the school yard, most of them hideous and all of them deformed, so it was with a puzzled frown he was finally able to make the various parts of the school rifle instructor.

"Me Sir? Yes Sir" Albert scuttled over. Oddly, the young Einstein was dressed in a black tartan kilt with a row of buttons the length of his black waistcoat, crested by a large floppy black necktie that bobbed like a crow clutching at his throat as he ran.

"You new here, aren't you?" Queried the malformed Cyclops, whose breath smelled sweetly of garlic and saltpetre.

"Err - yes Sir!" returned our intrepid kindernaut, not sure which part of the rifle instructor he was supposed to stare at, indeed if he was supposed to stare.

"Are you here for rifle or machine gun practice?"

"Machine gun?" An odd glow appeared in the young boy's eyes.

"Yes, little one, we just acquired a model of the Maxim water cooled, recoil-operated, three hundred rounds per minute machine gun. The toast of German technology."

"Machine gun?" The glow grew and spread across the face.

"Yes. Are you slow or just French?"

The firing range was further to the back of the school, originally designed as a conservatorium, but after due consideration it was decided that Wagner was so similar to the clamour of war, that any change of syllabus was one of mode and not of intent. In the midst of the hall, a brand new Maxim machine gun, Mark Four, water cooled, recoil-operated, three hundred rounds per minute machine gun was installed and duly aimed at a towering bank of sandbags where the orchestra pit originally resided, upon which a crudely picture of the Tsar of Ruthenia was fixed.

At this precise moment of time, Wilhelm Friedrich Ludwig III, Kaiser of the German Reich had arrived at the adjoining munitions factory. He was a stout Prussian gentleman with imperial moustache, well known for his kindness to children, being flawlessly polite, and his devotion to the rule of Prussian law and absolute war.

The young Albert stared in amazement at the riddle of barrels, water jacket and shining cartridges laying strewn about the Maxim gun; it was mounted on a two-wheeled carriage; with a seat on the yoke, not unlike a gun carriage.

"Sit down and assume the firing position", barked the Rifle instructor and school concierge, clapping Albert on the shoulders and knocking him to the gun-seat. "Now, ja ... but try to use the toilet next time you have the need. Now, pull back the arming lever and cock the ... no, your hand is stuck in the magazine feed ... ja ... you have nice buttons ... now, pull back the arming ... no, you may not suck your thumb ... yes, it is bleeding but you are ... no, do not pull that lever - that releases the wheel-lock ... no, put that back ... Gott! Not yet!"

As gun fired the carriage swiftly rolled backwards, and a spew of bullets tore out of the Maxim gun, at a steady rate of three hundred rounds per minute driving instructor and student across the school yard at a steady rate of eight metres an minute. "Vatter im Himmel!! Pull the brake!" The rifle instructor yelped out as the carriage rolled over him, snapping the wires in his wooden leg. "Cease Fire!"

It was to no avail, young Albert had frozen in fear and was hanging on for dear life, and amazement as the bullets tore off the heads of serried rows of flowers that surrounded the schoolyard. The gun cart skewed to the left leaving a neat tracer line of bullet holes along the side of the borders dormitory, before breaking free and launching a salvo that riddled the walls of the adjoining state munitions factory.

"COSSACKS!" was the universal cry of Kaiser Wilhelm Friedrich Ludwig III, and his attending officers dove to the ground amidst exploding cordite and phosphorus.

A short time afterwards, a very contrite headmistress was sent to head a missionary leper colony on Lake Tanganyika in German East Africa, the Einsteins were held in internal exile in a dungeon in Swabia for a short period, and Albert had his bicycle confiscated.

Einstein and the Duel

Courage can exist in even a dormouse, but doesn't. - Einstein

"You there, sir! Yes, you sir!"

The teenage Einstein looked around trying to identify who was old enough in the immediate vicinity to qualify as a 'sir'. Given it was a school yard and everybody there was wearing lederhosen and playing catch, it presented an undeniable quandary.

"You sir! I am talking to you!" Came a thin reedy voice that was punctuated with the occasional hiccup of a rapidly approaching adolescence.

Albert pointed to himself and looked at the other boy. "Me?"

"Am I looking at the Moon?" the boy said sarcastically, "Do my eyes appear cross-eyed? I am looking at you, sir, therefore I am referring to you, sir. Hence you are the Sir I am speaking to. Yes, you."

"Hello?" Einstein replied reasonably, "No, my friend, but you see I mistaking me for an adult. Oh, no, like you I'm still thirteen …"

"Do you mock me, Sir?"

Einstein decided to take a closer look at his inquisitor. He was a small boy with a cowlick for a haircut and snub nose, his blonde hair and pale blue eyes with their unblinking menace was unnerving, if not outright threatening. In his hand he held a stick or bat, and in his other hand he held a ball. When combined these were used for playing a German form of stickball. Einstein had the feeling that a rudimentary form of cricket was not what those washed blue eyes were thinking.

"I…"

"You're standing on the spot."

Einstein looked down and noticed for the first time the outline of a square where a batter was supposed to stand.

"Oh right …"

"I challenge you to a duel."

Immediately the chorus of 'Fight! Fight! Fight!' erupted around them as the other schoolboys formed a teen amphitheatre, one of those bloody Roman circuses that spontaneously come into being at the drop of a schoolbook.

The world of the child is not the universe of the adult. It is a closed, incomprehensible nation with a rule-set that shifts from moment to moment with a fluidity that rivals boiling mercury, where allegiances and empires come and go from one lunch break to the next. It is not possible to know the mind of a child dictator, only the bloody aftermath of another battle. In the world of a nineteenth century German schoolboy, fuelled on the stories of the Franco-Prussian war and an environment of competitive duelling, this was only more so. Duelling in the German Empire wasn't simply a pastime for slightly post pubescent men with hankering to be noticed. On the contrary it affected the entire mindset of the German peoples from the Emperor down to woodsmen carving dolls in the Black Forest. It was an unstoppable social force that swept the countryside and cites, rearranging more faces and stopping more lives than tuberculoses and syphilis combined.

"But … but …" the young Einstein protested, "but why?"

"Why?"

This brought the whole pack of screaming boys to a halt, the possibility of 'why' had simply never entered their mental space before, no one had ever actually thought about duelling, it was simply part of the ritual of school yard.

"Why?" the pale blue eyed boy blinked, "… cause you are standing on the spot, that's why."

Einstein stepped to one side.

"No I'm not."

"That's cheating!"

Einstein stood on one leg, to which the crowd of boys stared curiously.

"Why are you doing that?" asked the pale-eyed bully.

"I'm getting you to think about why you want to fight."

This was too much for the bully who lifted the bat, but Einstein was too fast and ran off with a schoolyard in pursuit.

Ten minutes later he found himself hiding in the school bell tower. He peered down like a fearful sparrow at the flocks of boys wheeling about the grass. His school years had been endlessly chaotic, one painful incident followed by another. He had become so adept at hiding that teachers had trouble knowing if he attended for one entire year. The problem was he was far too clever. The other problem was his fellow students were clever enough to realise this, and knew the best way to deal with him was to hit him often and to hit him hard. It only pays to stand out when you are not with your peers; otherwise they find ways to internally exile you to an adolescent Siberia.

Boys aren't cruel, they're simply pre-programmed by evolution to maim, destroy and kill - Mother Nature only nurtures success.

Take a sensitive, innocent and inquisitive mind, then thrust it into a belligerent and protocol obsessed society, which measures the value of a person on how well they can dominate others and you will have the history of someone who is tormented from day one at kindergarten. Each day seemed to bring new terrors. Failing to pay attention in German grammar class would result in an instant caning. Failure to march between classes happily singing resulted in a boot from behind. Sitting at the desk studiously doing one's arithmetic occasioned a rolled up newspaper over the back of the head. This last one was one of the most irritating aspects of the classroom, mostly because he simply couldn't understand its cause, which was – and he only understood this many years later – that some teachers are sadistic bastards.

Choir practise was obscene.

Sport was a bewildering array of different forms of corporal punishment, indeed it could hardly be called sport, as it was so orientated towards humiliation and pain that even going on the field could almost be considered preparation for battle. Perhaps that was the design of the teachers when they made up the games. The boys didn't simply play hockey with their sticks, it was more akin to violent fencing with the occasional consideration for the ball and even rarer thought for the goals; so many boys would end up with gashed scalps and chipped shins that no match had ever made it past halftime without the entire squad being invalided. Very early on, the young Einstein had mastered the art of having an asthma attack almost immediately after going on the field. This had the peculiar flow-on effect of his developing extraordinary powerful lungs, which in turn helped in his constant escaping from lynch mobs of raving juveniles.

Suffice to say, he had an endlessly miserable time and hated every moment of it.

Yet, it was the duelling that caused him the worst suffering. Duelling had become a social madness throughout the continent, except in Italy where everyone always showed up too late to fight having spent their entire morning dressing to die.

In Germany particularly, a man's need to show he had not only fought a duel but also survived was paramount. If by the age of graduation from university a young man had not gained a facial scar, he was brought before the school's bursar and asked if he was an

Italian. It was, of course, completely illegal. That didn't matter and only added spice to a man's reputation at his having broken the laws of the country as well. After all, if two young men and their seconds at arms, wish with all their hearts to meet in some abandoned place at daybreak and proceed to stab, slash or drill bullet holes in each other, there is very little the constable of a sleepy village will do about it or even bother.

The stories of these battles were all relayed to an eager public starved for entertainment between wars, whether the newspaper was the Berliner Gazette, the Illustrated London News, Le Temps of Paris or even the Budapest Yiddish Chronicle. A successful encounter and it's expurgated glory on page five could almost guarantee a young man a diplomatic post or high ranking position in an army. If the duel was famous enough – it could be any army.

For the teenage boys at Einstein's school, duelling took the form of the Wall. Sixteen feet of sheer terror over the canal, with broken glass embedded in its top, on one side the drainage from the local knackery filled with dead horse heads and their intestines and on the other an unobstructed view of the graveyard as it trailed up the hillside and disappeared into the darkling woods. The object of this duel was to remain on the Wall, as along as possible without retreating, falling off, or – the Kaiser forbid – surrendering to one's opponent, as the spectators would dance about on the gravestones, hooting encouragement for their champions and booing the inevitable loser.

One marvellous battle lasted over nine hours, until the full moon had risen and everybody else had gotten hungry and gone home. It was a candle against which all subsequent battles were viewed. The two boys eventually agreed on a draw and carved both their names on the top of the Wall, which was the traditional honour awarded to the winner – normally the spectators would have thrown the loser in the canal, if he hadn't already fallen.

The unwritten rule of the duelling on the Wall – or was it written in teenage blood –that the boys allowed themselves only one weapon each, and these were whatever they brought with them; consisting of sticks, poles, rocks or even the fabulous air-rifle. In a sense, the air rifle was considered a cheat as there was little or no defence against it, barring an all out charge across the parapet and knocking down the gun-boy with a stick. Thus, in the same way the game 'paper, rock, scissors' could be won with a foreknowledge of a foes mind, so knowing which weapon your opposing duellist was carrying could mean the difference between hours of agony and a trip to the doctor, or being carried around the mausoleums as the days champion.

The best way to win was to simply march across the rampart and hit the opponent as hard as possible and as often as possible until by sheer weight of power the hapless boy was driven down, out or off. The broken glass, however, brought special dangers as one slip might bring an awful gash to the knees and required either very strong boots or the artistic ability of a male choreographer from the Bolshoi Ballet, and more then one hero had prematurely lost a battle due to cheaply shod feet – boys with clogs had a special advantage.

Any form of defence, be it shield, helmet or mother was reviled.

One boy had once turned up with an ancient blunderbuss with its terrifying flaring muzzle. He wasn't able to secure any black powder, but the sheer size of its enormous barrel was too much for his challenger, who upon reflection decided he really was too young to die and quickly beat a retreat.

That evening Einstein addressed his parents over the dinner table about the problem.

"I'm going to die."

"Nonsense, we have already paid for this term at school. They don't give a refund, you know." His father said emphatically over his dumplings.

"No, papa, really I am going to die!"

"Listen to your father," said his mother, "he knows about dying, he's says his been dying from my cooking for the last thirty years, and still nothing has happened. He's an expert in these things. Now eat your food."

His father merely grunted at this one.

"I don't think you understand, I've been challenge to a duel, with the most dangerous boy at school. I really could die."

"Well, don't go, here I'll write you a note."

"Mama, it's not something you can fix with a note."

"Of course it is, remember that time … no, wait that was your uncle Sammy… and didn't he go to prison for that?"

Einstein the younger stared helplessly at his food.

"Hmm, a duel you say," his father finally mused between mouthfuls.

"Yes, papa."

"Chutzpah."

"I thought it was dumplings," the younger toyed with his food.

"No, I meant use some chutzpah. Just brazen it out. If you're going to die anyway, might as well just bluff your way out, after all, you have nothing to lose."

"My life?"

"Trust me, my boy, you have nothing to lose."

Thus it was one spring morning the adolescent Einstein was challenged by the most dangerous boy in the school to battle the next day:

"You there, sir! Yes you sir!" the same reedy voice piped across the playground with the dramatic tension of Achilles challenging Hector at the Siege of Troy.

This time Albert found his exit was blocked by the fourth form sports master watching sternly from the sidelines.

"I think he is addressing you, Einstein," the master said, having an inkling of what was about to unfold.

"Me?" Einstein looked around bewildered as his intestines wrapped themselves around his lungs.

"Yes, sir, you sir." This, all coming from a boy who had just discovered the joys of popping his own pimples, was a little remarkable. But then again, Manfred von Richthofen was a remarkable person, and one day as Baron von Richthofen would become world renown as an extraordinary World War One fighter ace. At this point in time, however, he was only three years younger than Einstein, but this didn't seem to faze him in the least, if anything, it made him into a writhing ball of cold fury.

"Again?" Einstein asked.

"What do you mean again?"

"Ah, nothing."

"I challenge you, sir, to a duel!"

"What for, this time?"

There was a short pause as Richthofen thought of an acceptable insult.

"My shadow! You are standing 'near' my shadow."

"Near? Near? Well, yes, but I'm not actually standing on it. I am a whole pace to one side - see?"

Manfred stepped to one side.

"Now you are," he said coldly.

By this time, the fourth form master had disappeared on his

rounds and there was no hope of official condemnation. 'It was on.' – semantically the adolescent Einstein knew it made no sense, but who said adolescents had to make sense. At that moment, the after-school bell rang and there really was no reprise, as both boys marched at the head of the school yard to the dreaded Wall.

Thus began one of the most extraordinary duels that the Karl Friedrich Strasse Hochschule was ever to witness.

"Wait!" Einstein cried out as the boys assembled themselves and had scattered across the tombstones beneath the Wall.

The assembled hooligans all looked at this indecision with disgust.

"Are you forfeiting?" asked Manfred von Richthofen with a cold sneer.

"Err, no, not at all, I merely wish to consider the legal and historical ramifications of this duel within the context of the struggle of the democratic process in a post Napoleonic era."

"What?"

"Err, not buying that one," Einstein flustered.

"Get on with it," came the cry of a dozen bloodthirsty boys.

"Wait!"

"Well?"

"You challenged me, right?"

"And?"

"That means I get to choose the weapons! As set down in the Irish Rules of Duelling of 1777!"

"Nonsense!"

"Here see," and surprisingly produced a copy with the law underlined.

"Damn! That's not how we do it!"

"Rule book say otherwise."

"Alright, so choose then!"

"I choose, ... I choose..."

"Just choose!"

"Insults!"

"What?"

"Insults at twenty paces!"

"You can't fight a duel with insults!"

"Yes, you can, says so here, see," holding up the pamphlet again and read the section aloud, " – the weapons of choice are to be decided by the person who is challenged. That's me, so I choose combat by slander."

This brought the whole proceedings to a halt, which was Einstein's intention, and they spent the next hour arguing whether it was technically feasible to have a fight to the blood with nothing more than teenage witticisms. In the end they had to agree that while he was right since insults were meant to be 'cutting', he was certainly no fun.

"Alright then," Richthofen finally agreed, a little more relieved then he would not admit, "we fight with insults."

"Also," Einstein continued, "no insults about mothers or knobbly knees."

Mothers they could understand, but knobbly knees were ridiculous, so they voted against it. This too, was Einstein's intention as he didn't want his mother to find out he couldn't defend her. Part of his mind also wished he had chosen his mother's cooking, it did after all have a certain lethal aspect to it.

They clambered onto the Wall, this aspect of the battle was not negotiable, after all, the entire crowd wanted to see blood, and by the Kaiser they were going to get it. Einstein due to a history having his mother's dumplings every night for the last thirteen years did have some trouble, as his body had taken on a geometry not unlike a dumpling, but he succeeded eventually. He felt an asthmatic attack coming on, then looked down and shuddered at sight of the former head of a draft horse called Boxer float past. Since they would throw him in the canal if he did lose, losing wasn't an option.

His only hope was the tried and true tactic of outtalking his opponent, not so much he infuriated him that it disintegrated into an actual fight, but just enough to let him win on points. Picking his way across the broken glass cemented into the rampart, he noticed one patch was covered with slimy moss, he stopped just before that and waited.

Meanwhile, the younger and almost a head shorter Richthofen had bounced onto his end of the Wall with the enthusiasm of a polo champion winning a chukker and standing on his horse to celebrate as it pranced about the field. This was his first duel and he had chosen his opponent with care, Einstein was obviously bigger than himself, but so timid and clumsy that Richthofen was confident of winning. The problem of the duel turning to a slanging match made him even more confident, yet he too noticed the turgid remains of Boxer the horse float past and managed to keep a tight rein on his enthusiasm lest he be defeated by his own hubris. Not surprisingly this same sangfroid would protect him through endless battles above the fields of the Somme in World War I.

"Are the combatants ready?" asked the senior boy, who was the unofficial judge of the contest.

"Um."

"Yes, we are ready."

"Before, you commence I must ask you, if honour will be satisfied by shaking hands?"

"Yes, oh yes!!"

"Never!"

"So be it!"

"I said yes!"

"According to the rules, the agreement must be made by both sides," then the senior boy grinned, "unless of course, you wish to forfeit and accept the penalty for losing."

"Let me think about,"

"Let the contest begin," the senior boy dropped his handkerchief and the chanting began. The chanting was part and parcel of the fight, giving it an atmosphere not unlike the more dramatic bars of Carmina Burana.

"FIGHT! FIGHT! FIGHT!"

Einstein would have fainted except he was too afraid to, and it was only this point did he realize he hadn't a clue as to how to insult someone at close quarters and not be torn to pieces.

"You ... you're ... Short!"

"That's not an insult," countered Richthofen, "I actually am shorter than you. No point."

"No point!" cried out the senior gleefully.

"FIGHT! FIGHT!"

"Aw!" and Einstein found himself staring once more at the head of the dead horse.

"Challenger's right of reply," continued the impromptu judge.

"You play hockey like a girl!"

"Hooray!"

"Point!"

"No, wait! - I, I really do play hockey like a girl!"

"No point!"

"Oh come on!"

"If he agrees, then it's true and no point."

"Who made up these rules?"

"I did, as senior prefect, you have a problem with that?"

"Damn it!"

"Defender's call."

"You are a girl!"

Richthofen ground his teeth in frustration; since this was something no boy could possibly agree to.

"Point!"

"You kiss boys!"

"No, I never!" Shrieked Einstein.

"Point! One Point each!"

"You're wearing your sister's underwear."

Oddly enough, Richthofen had been forced to wear his sister's underwear that morning, but who in their right mind would agree to show it.

"Point!"

"Then you are your sister!" He riposted.

"That makes no sense at all!" Einstein contradicted.

"None of this makes any sense!"

"Point! Two points each, and get off sisters, already," the senior prefect ordered.

"A hedgehog has a better haircut than you!"

They all thought about this one, it wasn't so much an insult as an observation, so plastered down with lard was Richthofen's hair.

"No Point!"

"You've got a head shaped like a toilet!"

"Objection!" cried Einstein.

"Just tell us what the objection is," the prefect said, enjoying himself immensely, "let us decide if it actually is an objection."

"I got my head from my mother! And we agreed no maternal insults!"

"Oh come on!" Richthofen cried and would have stamped his foot in frustration but remembered in time how dangerous it was on top of the Wall.

"Well," the senior boy mused, "technically he didn't actually mention your mother."

"But he did infer!"

"Insult by inference, hmm, alright no point! Still two points each!"

"Gah!"

The decision did, however, bring murmurs of protest from the bystanders.

"Hey! I am the head-prefect! I am authorized to hand out detentions, you know!?"

This brought a moment of quiet reflection.

"Where are we," said the head-prefect, "oh right, go on fat-boy."

"Hey!" said Einstein.

"As you said a truthful observation isn't an insult," the head-prefect chortled in his element, "so on with the show, lard-brain!"

Einstein flustered.

"Hesitation!" Richthofen almost yelped.

"So?"

"Well, I was just pointing it out."

"Get on with it!"

"You, you…" Einstein started.

"Repetition!"

"Which word competition are we playing here?"

"I'm sorry I haven't a clue as to the rules!" Richthofen explained.

"Let me decide, okay?" The judge frowned. "This is a blood sport, remember?"

In that time, however, Einstein had come up with a zinger.

"You're a teacher's pet!!"

"POINT! POINT! POINT!" went up the cry around the grave-hoppers.

"Yes," mused the prefect, "definitely a point that one."

The tension was getting even to the steely-nerved Richthofen.

"I ... I ..."

"Repetition and hesitation!" cried out Einstein.

"I thought we had covered this already?" said the prefect.

"I can hit harder than you can!" Richthofen screamed out triumphantly.

"FIGHT! FIGHT! FIGHT!"

The head prefect however, questioned the legitimacy of this retort.

"If you do hit him, than that would be outside the agreed method of combat, on the other hand, if you don't hit him then we can never know if it's true or not. So, no point! The battle is over; Einstein has scored three points, Richthofen two. Einstein is the winner!"

"No!"

"Will you jump? Or must you be thrown?"

"No!!!"

"Into the canal with him, men!"

The tension finally broken as the crowd swept forward and dragged Richthofen off the wall, and carried him down to the weir, and threw him in as they sang Handel's - "All hail, the conquering hero!"

Einstein and the prefect stood behind them, watching from a distance.

"Hopefully, they will never figure out who you really are," Albert grinned at the senior.

"Yes, cause in that case, we will both be in the canal," agreed his elder brother as they picked up their satchels and walked home.

Newton and the Case of the Masons

The Enlightenment was a dark place. - I. Newton

A single giant chandelier illuminated the reading hall of the Gentleman's Club of Spalding. It was a voluminous room, steeped in a two hundred year tradition of decorum and inebriation, and in this blessed institution its members steeped themselves with as much inebriation as their gout would allow. Its chairs were constructed of rich leather, high backs and great privacy, the sort of seclusion where a fellow might disappear for hours on end with only a newspaper and port to succour them, some of the elder fellows had disappeared for days before being discovered beneath a farming gazette. The more energetic and sprightlier septuagenarians made occasional forays to the billiard table for conversation, and it was here that one afternoon Newton found himself in the company of one of the more unusual members of the association.

Charles Bartholomew Netherby was a tall and florid clergyman given to espousing upon subjects of the most trivial details with endless enthusiasm, much to the annoyance of whomever he managed to buttonhole.

"Butterflies are really the most extraordinary insect," said Netherby as he polished his billiard stick and peered over his spectacles at the table. In the dim light of the room his red face was vividly offset by black waistcoat upon which dangled a bright Masonic badge.

"Fascinating," said Newton with uttermost boredom, "it's getting late."

At the Gentleman's Club of Spalding it was always three o'clock in the afternoon, regardless of the time or the year or the actual time of the day. This was an article of faith set down many years before in the clubs minutes by a spelling error, which had never been rectified. The members, however, had discovered this made appointments and meals infinitely more easy to manage, as any time was the right time and the club was of course the right place, provided they never left the building. There were three general topics of conversation in the Gentleman's Club of Spalding, last week's foxhunt, how much they had lost at gambling, and the size of the bullet holes they had obtained in the Civil War. Scandals were common in the club, stemming often as not as someone sitting in the wrong chair or reading the wrong newspaper.

"I noticed this the other day when I came across a story of that Thomas Blood."

"Thomas Blood" said Newton, "is a man who has boldly run through more villainies than England ever knew. A more vile highwayman could not be found."

"You know him?"

"Sir, I arrested him," Newton said triumphantly, "it was I who sent him to the gallows."

"Odd," mused Netherby as he sank another ball, "he seemed remarkably sprightly for a dead man."

"What? You must be mistaken, Thomas Blood was taken to the Assizes and hung not two years since."

"My dear fellow, I assure you I have met this Master Blood and he is very much alive."

Newton dropped his billiard cue.

"Damn his eyes! The man escaped!"

Newton ran out the room and clattered away. Netherby grinned and pocketed their wager of two sovereigns lying on the table.

"Works every time," he smirked.

Newton returned two days later and confronted Netherby as he was setting up the billiard table.

"Damn your eyes Netherby!" yelled Newton, "Thomas Blood is dead. I returned to London and had his body disinterred, and I can state for a fact the man has being decomposing for at least two years. I know this by the state of the body."

"Really?" Netherby asked mildly, "hmm, must have been a different Thomas Blood. Care for another game?"

"Is that all you have to say?"

"Well Isaac, you can't blame me for you mistaking your Thomas Blood for my Thomas Blood. I never said they were the same person, I merely said I had met Thomas Blood. Two sovereigns?"

"What? Blast, damn somebody's eyes. Oh very well. Two sovereigns."

Netherby broke and soon confiscated half the balls.

"I remember now, I've seen that tomb, there was something odd though," said Netherby.

"What was that, old man?"

"The tomb of Thomas Blood, which is at St. Margaret's Church near Saint James Park, it was very richly endowed, indeed one might almost call it a mausoleum."

"That would be Sir Thomas Blood, late of Norfolk. Fought on the side of the cavaliers, I seem to remember."

"On the contrary, this fellow was the same to whom I have referred." Newton potted three balls in quick succession. "I know this, because of the date of entombment match that of the villain. I assure you he was no nobleman."

"Thomas Blood, born 1618 - August 24, died 1680?"

"Yes, you seem well acquainted with the fellow."

"I should, he recently become a member here."

Newton miscued the white ball and in one of his rare mistakes it shot off the side of the table bounced off a wall, and hit one of the other members of the club in the head as he sat in his chair.

"Zounds!" came the cry of a fossil as he discovered a billiard ball in his London gazette.

"Bad shot Newton," remarked Netherby, "unusual for you. I haven't seen you make a miss like that since you lost seven hundred pounds on the Exchange following the South Sea Bubble."

"Blood - a member of this club?"

"Still is."

"But that is simply impossible. The man is dead."

"He said he had taken the knighthood of Sir Blood, and arranged for his death certificate to match of the poor dead Bleeder, as we called him in the regiment."

Newton drove his pool cue so hard it buried itself in the opposite cushion, and the ball spun harmlessly lining up a shot for Netherby.

"Here?! When?"

"He boasted it to me, while you were away at St. Margaret's. It appears the fellow has been member ever since he escaped from the gallows. I believe he paid his way out."

"Paid his way out? From the gallows?" Newton almost had an apoplectic fit. "Here?"

"I suspect it was a way of getting back at you old man," Netherby looked smilingly at the ceiling.

"Guards!"

"We don't have any, try calling for the porters."

"Porters!" Newton screamed and ran out of the room as once more Netherby grinned and the two sovereigns into his waistcoat pocket.

Ten minutes later Newton returned triumphantly to the reading room, with the head porter in trail carrying the club registrar.

"Netherby you dolt!" Newton slammed the book on the billiard table. "There is no Thomas Blood here, and the head porter says he has never even heard of him."

Netherby looked over the book and then pointed out a name.

"Thomas Budd," he put his finger on it, "There, Sir Thomas Budd."

"Budd! I thought you said Blood."

"You did? Great heavens, what a mix-up."

"That is putting it mildly, mister Netherby, good day to you."

"Same time next, Newton?"

"Damn your eyes!" Newton yelled as he slammed the door.

The head porter looked mournfully at Netherby and closed the book.

"Brother," he said with a tone of deep reproach, "you really shouldn't tease him so. It gave me the world of fright."

"Well Thomas, I don't know, things would be so boring if I didn't."

Einstein and his Moustache

Shaving is a Science. - Einstein's Dad

One morning, as the sun rose dull and unrewarded by the clouds, our hapless hero the unremarkable young Einstein, woke up and thought about his daily chores of achieving nothing more than washing his head, cleaning behind his ears and wondering which bicycle to ride.

Berlin roused itself from its winter slumber, a brown slush of mud plastered the streets, leaving its inhabitants to grumble about unhygienic conditions and cursing run-away carriages. Nobody was cheerful, anymore than the folk of Berlin were allowed to be cheerful, for Berlin was a city of rules and council regulations dealing with just about everything. From walking the dog, cleaning up after the dog, walking the children, cleaning up after the children, painting the front of the their houses to match the front of all the other houses, painting the children to match the dog, and so on.

Einstein at this point still lived with his parents, and was still as unexceptional as everyone else, except this morning, when he stared in the mirror and noticed for the first time, a wrinkle of hair upon his upper lip.

"Ach!" He exclaimed in dismay, "I'm could go to prison for this one."

It should be noted dear reader, that the state law and regulations pertaining to the growth of hair were explicit on this point. For the purposes of identification, a person, above the age of enlistment into the Wehrmacht and below the age of consent, was required to give due notice to the respective authorities for the possession, growth or removal of any hirsute feature upon the face.

Or, more simply he had to tell the police he was growing a moustache.

He rummaged through the bathroom cabinet, knocking out patent cures for dandruff and familial lice, finding nothing more applicable than a depilation crème for warts.

"Papa! Papa!" He cried down the hallway to the kitchen. "Where is your razor?"

"I have a beard!" Came a mouthful of eggs and toast.

"Is it in the beard?"

"I don't shave, you son of a potato!"

"Oh, yes."

Albert stared back into the mirror; this was going to be a hard one.

He read the instructions for the wart remover and carefully applied the crème to his upper lip.

"Mein Gott!"

"Have you found a razor?" Came the voice of a cautiously optimistic father.

Einstein hopped around the bath for a moment rubbing his lip then shoved his face in the sink to wash off the wart blistering crème. He peered worryingly into the mirror, a bright rash was apparent beneath the nascent moustache, but the hair itself was untroubled by it's adventure. He dried his face off and wandered into the kitchen, where he father was onto his third cup of coffee.

"Papa, I am a man."

His father looked up with disbelieving eyes.

"Have you a job?"

"No, Papa, I have a moustache."

"That doesn't count. It's more of a fungus, and why is your lip so red?"

Albert explained his novel use of wart remover.

"My boy, you are in that case, most definitely not a man. Now go see a barber."

"Yes Papa. But I need a permit."

"Ah, you should have thought of that before you attained puberty shouldn't have you, and what you have on your face doesn't warrant a permit. Now off with you boy," and waved his hand idly, pleased that his paternal gifts were not useless.

"Papa."

"Yes?"

"Does this mean I can have my bar mitzvah?"

"Hmm. How old are you?"

"Nineteen."

"Nineteen?!"

"Yes Papa."

"How did we let you get to nineteen without having a bar mitzvah?"

His mother looked over her newspaper. "He had the measles that week, we all went to Paris and left him behind to get over it."

"Yes, I remember now," The pater reminiscenced, "it was gorgeous weather down there. Paris in the spring; April in Paris," he trilled, "Ah, and so, so, very quiet. My boy, you really must get measles more often, for the good of the family."

"Yes Papa."

"He can't." His mother opined.

"Can't what?"

"Get it twice. He's immune now. You can only get measles once. He's almost indestructible now, from the point of view of a measles virus, that is," somehow conducting the sentence with a cup of coffee. "Measles germs look upon his immune system in the same way we look up mount Everest, it can't be down they say."

"Measles have conversations?" Came the startled looked of

the father.

"That was irony, now eat some more eggs and toast. And you," pointing at the boy with the moustache, "go see the English barber." His mother retreated behind her newspaper.

"Mr Todd?"

"None other." Came a distant reply. "He gives cut rate prices"

Head swaddled in a towel to hide his moustache, Einstein made his way along the streams of mud, beneath the unrelieved soot of German Imperial buildings. He had barely gotten anywhere when a flock of children dove and wheeled about him, like mudlarks picking moths out of the air.

"Elephant! Elephant!" They cried gleefully and danced in a circle around him.

Albert looked around seeking the elephants, before it dawned on him that he was the object of their song and the aforesaid pachyderm, he stopped and pulled off the towel to ask them what they meant.

"But I'm not an elephant, here, look."

The urchins, one and all, peered up and expressed annoyance at being lead on by him, firstly for his not being John Merrick - the Elephant man, they picked handfuls of mud and began pelting him, and secondly for being the odd guy who lived up the road. They had been lead to believe by posters about the town that a carnival show of the esteemed John Merrick was coming, and that this was indeed the marvellous mister Merrick.

Einstein ran off, wrapping the towel once more about his nascent moustache, crying, "I am not an elephant! I am an Einstein!"

It was this extraordinary sight that charged down the street, passing none other than the aforementioned John Merrick, who was riding a carriage to the Berlin institute of Medical science and

Abnormities. He shouted out from the carriage.

"I AM NOT AN EINSTEIN!"

Looking up at the luminous lettering of the barber sign, Einstein found himself having some misgivings. It wasn't the gothic gold script that flourished proudly the name 'Sweeny Todd – Barber, Dentist, Doctor, Surgeon, Wife Remover' that worried him: it was the scratch marks on it's side as one entered the door.

"Come in, come in dear boy, take a seat," a moon faced heavily be-whiskered barber said to him. Clumps of hair splattered in blood stuck to his apron. "I've just got to removed this tooth, won't be a moment, now take a seat. Here, read the London Illustrated News."

A man was strapped to the seat, moaning and kicking with his feet, with a brace holding his mouth in place against the headrest. In the mirror Albert could see his wild eyes staring at him, seeming to communicate a plea, but Albert was able to ignore the issue by turning to the comic section in the newspaper from England.

"There now, how are we, hmm, wont' be long, just five more teeth and I'll be able to remove your tonsils. Hmm? What's that? Oh, dear fellow, I'm all out of pain killer, here crème toffee?"

This lodged in the poor fellow windpipe and with a burping cough it catapulted out of his mouth and bounced off the mirror.

"Allergic?" Quizzed the barber, "Bother, wished you told me, crème toffee are my favourite."

Einstein sank further into his chair and deeper into the newspaper of the London Illustrated News. With its intaglios from the Franco-Prussia war, enormous cannons flaring over the ramparts of Paris, as troops held poses that seemed more in line with playing charades than destroying an empire. There were letters to an editor so indifferent to their needs they ended up sandwiched between the section for lonely hearts and novel patented medicines for leprosy, insomnia and hysterical children. Oddly all the medicines seemed to have the essential ingredients of opium and gin. It was while pursing these adverts that he came across a notice for public attention, pertaining to the whereabouts of a certain barber from Fleet street who went by the name of Sweeney Todd, wanted in connection with the mysterious disappearances of certain gentlemen and their butlers.

There was a yelp of pain as Todd drew forth another tooth from client.

Einstein looked up and stared at the large back of the barber, hairy arms bulging with muscles poked out from an enormous torso, the sides of the apron struggled to contain a bloated body which fell in clump to broad stripped pants, which in turn narrowed to his tiny dancing feet, and about the waist hung an unmistakable butchers knife, sheathless it swung ominously keeping a rhythm with the barber's mincing step.

Another yelp, another tooth tossed into the basin.

Einstein started shuffling to the door, holding his newspaper in front of him for cover.

"Now don't go anywhere, young fellow, I won't be long." The barber's bulbous head stared at him through the mirror, holding up in one hand a bloody forceps and the other gripping the tongue of the patient. The adolescent retreated to his chair, quivering behind a slanderous review of Oscar Wilde's - Lady Windermere's Fan.

A final shudder came from the customer on the barber's chair before he fainted.

"Much better," said Todd, letting go the client's tongue and talking to Einstein, "some days they never faint, and I'm here for hours. All this fussing and fighting, I mean do they want the job done or not." He shook the blood off his fingers. "Now then, young fellow, while we have this moment of peace, how can I be of assistance? Haircut? Tonsils? Vasectomy?"

Einstein gulped air, not unlike a frog trying to decide if it wanted to be a mammal or a fish.

"Yes?" The wide beamish face of Sweeney Todd loomed a few inches from Einstein's equally pale as the moon face.

In a moment of inspiration: "Cig...cig..." he attempted cigarettes.

"Speak up lad, I've got a load of sweetmeats to unload before it gets dark."

"Cigars!"

"Oh, that's all," he reached to the shelf above Einstein and brought down a cheap packet of Dominican Fuentos. "Two pfennigs. Anything else?" Pocketing the change.

Einstein held the novel means of escape in his hand and bolted toward the door.

"Hold up." A grim hand grabbed him around the collar and put him back on the chair. The Barber opened the packet of cigars, stuck one in Einstein's mouth and lit it for him. "There now. The other lads will take you for a real man now. Off with you."

Gasping for breath, Albert leant against a wall, his eyes bulging with terror.

"Permit."

"Wha ... "cough, "what?"

"Permit."

Albert followed a line of startlingly bright buttons, to a chin chiselled by Thor's hammer and a granite face surmounted by a patented leather hat so polished it stopped traffic. Staring at the police officer, Albert thought he could almost hear a martial band marching along the street, and then to his great surprise saw there was in fact a military band stamping it feets, lead by the officer.

"I … I," Albert began and then held his finger against his top lip; this brought the cigar back into his mouth and caused another coughing fit. "I don't have one."

The marching band waltzed its way down the street, waved on by the police officer who decided to work his terror upon Albert.

"You must come with me to the police station."

"Name."

"Einstein."

The police Sergeant leant across his desk and stared down at the young offender, and made a note that took five minutes and three inkwells to finish.

"Ah, you're the Einstein boy aren't you?"

"Yes sir, I have that affliction."

"Right then, here," and passed a scroll of yellowing paper to him then made another five minute and three inkwells note. "You're still here?"

"Yes sir," Albert hesitated.

"Well?"

"Sir, um, what do I .. um?"

"That's your permit, you may go. Now be off and don't come back till you've broken a law or joined the communist party."

"Yes sir!" Albert saluted and clattered out the front door and ran all the way home, dodging the children throwing lumps of mud and knocking down a medical curiosity.

He charged into the kitchen where now his parents were halfway through lunch and thinking about tea.

"Papa! Mama! I'm a man … again!"

"Did you get a shave?"

"No, a license to smoke!" And puffed excitedly on his newfound habit.

Einstein and his Passport

Stevie Nicks was no gypsy – Einstein

"And bring back a cuckoo clock!" Albert's father yelled out to him as he walked out the door. "Not a tram like last time!"

"Every cloud has a silver lining," Albert yelled back.

"Where?" his father shouted angrily.

"Oh .. you know." Albert attempted nonplussed.

"And don't cross the bridge, till you've made sure the ferry isn't running!"

Einstein the junior, puzzled over the wisdom of that one, as he cycled away. His father, Einstein the Elder, upon finding the Younger lying about the house with nothing to do, had decided it was high time the Younger was off for a holiday and sent him off to the Black Forest to buy a Cuckoo clock, or anything to get him out of the house. The Younger mantled in his old duffel coat and moustache, coasted down the Berlin streets, for all the world a piece of human flotsam drifting through life. Above, the Linden trees were exploding in their autumn colours, an unbroken downpour of falling leaves covering the road with a river of decaying colours, and about him carts and pedestrians swam up and down the stream of glowing foliage. Einstein stopped, and looked up at the crimson storm of trees.

"Really, this weather is getting worse everyday," he muttered and forded the street filled with leaves.

The vacant face of a border guard, stared sightless into the driving leaves and across the empty fields, there too, broken trees were disgorging their summers feast, it was a sight that little moved him, for autumn was the worst season for guards with it's constant rain and endless leaves to rake up. To be a border guard brought no pleasure, his only real joy came from his wildly inaccurate misdirection's; extraordinary prognostications of geography were for him as works of art, his own individual answer to the brutal logic of the State Law and Regulations. If someone asked the way to Berlin, he would often send them to Budapest, or someone seeking to find the road to Düsseldorf would end up in Poland. In his own manner, he was the Van Gogh of Bureaucracy.

"Passport?" He barked at Albert, who had reached the edge of Brandenburg, the first border on his way to the Black Forest.

"Passport?" Quizzed our cyclic adventurer.

"Yes, a passport, you need a passport." This eternally weary sentinel stared unblinking at Einstein. "It is the state regulations that all persons wishing to travel across the internal borders of the German Reich are required to produce a passport - or a bribe," he glowered unenthusiastically and then winked.

Albert took a moment to digest this, then said, "Ah, I have forgotten both my passport and my bribe, I will go get them," turned his bicycle around and headed home.

The guard watched the retreating cyclist, grinned as knew he was in for some fun.

Back at his house, Einstein's father found him looking through his roll top bureau, papers and pipes lay strewn across the carpet and the furniture, a landscape of notes and scribblings decorated his study like the aftermath of the sack of Rome.

"Ach, father, I've found my passport, but I don't what my bribe looks like."

For several minutes there was a slow fuse burning, but the youthful Einstein was unaware who was burning. He looked up, wondering why his father was staring at him so intensely. There was a short explosion, as papa dragged him off to the front door, shoving his passport into pocket and handing him two Pfennigs before slamming the door behind him.

"Ah! It looks just like a Pfennig." Einstein the Younger observed then rode once more into the stream of autumnal leaves.

"Officer, here is my passport and my bribe," beamed our cyclic adventurer.

The guard looked over his shoulder, about the crossroads and back again, one could never assume the Custom's Inspector was not out riding. Opening the passport, he saw it was out of date and incorrectly stamped; this brought a sense of immense job satisfaction to the guard. He grinned jubilantly and took the bribe - it was becoming a grand day.

"Hmm ... destination?"

"What?"

"Where are you going? You are going somewhere aren't you?"

"Ah, oh yes, to Schönwald in the Black Forest in Baden-Württemburg, for a Cuckoo clock - they make the best you know."

Another mischievous grin from the guard.

"Have you a permit for a Cuckoo clock?"

This brought some rapid blinking from Einstein.

"A permit for a clock?"

"Yes, state regulation controlling the transport of Time Pieces across state borders, stipulate that all such time pieces, be they fob watch, mantle clocks, desk clock, clock towers, Cartel clock, grandfather clocks, hourglass, sundial or Cuckoo, be they driven by water, escarpment, spring, pendulum or weight, that all such chronometers must have accompanying permits. As signed by the authorities of the State they were manufactured in."

"For a Cuckoo clock?" a long drawn out stare.

"I don't make the customs laws, I merely impose them remorselessly."

"Um, where?"

"You must apply for a permit to transport chronometers from the bureau of State Affairs in Baden-Württemburg, where the said Cuckoo clocks are manufactured."

"Thank you, sir. Thank you." Albert sighed and mounted his bicycle, circled the crossroad twice, then turned back and decided to ask for directions.

"Passport."

"What ... but you just saw it."

"Passport."

"But ... but ...but ..." mouth agape.

"Passport."

"Oh for the love of Gott. Here." Albert brandished his passport once more.

"This is out of date. It is incorrectly stamped, and has no bribe attached. You may not enter the border."

"What?!!? But I just did!"

"No, you didn't"

"Oh, yes I did!"

"On the contrary, you did not, in fact you left the State of Brandenburg, and you did not enter. That is different. For this, you require a new passport."

"Oh! Well, where can I get a new one?"

"You may apply at the passport office in Berlin."

"But Berlin is in Brandenburg."

"An accurate observation."

"So, can I go and get one."

"For this you require a passport."

"No."

"Yes."

"But that's impossible."

"Yes. I've noticed."

"Can't you help me?"

The guard gleefully looked about the empty fields.

"Regulations for border guards for the German Reich stipulate that border guards only have jurisdiction over matters within the border of their State. In this case, the State of Brandenburg. You are in Saxony. I am unable to assist you."

Einstein felt a cold wind being to blow from the East, it brought whispers of Siberia and empty forests filled with wolves.

"But ... but ..."

"You could try asking Wilhelm, he is the guard for the Saxony border. He is about a kilometre down the road."

"Will he help me?"

"No, he is forbidden by the Customs Regulations from helping people who do not have the appropriate visa and passports."

"I was worried you would say that."

A sense of panic settled into Einstein stomach like cold sour strudels, as the distant wolves of Siberia caught a whiff of his fear. "Are you saying ... what are you saying?"

"That your passport is out of order, you may not enter the state of Brandenburg, and you must apply for a new one from a German embassy from outside the German Reich."

"Oh."

"Now then, just between me and you." winked the guard. "And I wouldn't want this getting around, after all, I have to follow the regulations and rules set out for border guards to the letter. Important job this, you know."

"Yes?"

"Well, you could try that way," the guard almost cackled with laughter as he pointed the way to Budapest.

The wolves in Siberia began to howl.

Three weeks later, a very tired Einstein entered Budapest, he had had innumerable flat tires, slept in barns and ditches; was lost for hours in blizzards of autumnal leaves; and been chased by dogs of every description including some he could have sworn were wolves. He dragged his bicycle into the northern part of the city and collapsed at a coffee table.

"Yes?" said the voice of a waiter surprisingly fast.

"Can I help you?" Came another voice.

"Coffee?' Followed by a third.

Albert looked up and beheld three brightly coloured waiters, with notebooks and pencils poised. They seemed identical copies of each other with their hair oiled flat across their heads, dangerously waxed moustaches and brilliant smiles, which scared him with their enthusiasms.

The coffee shop was a vast auditorium to the art of the barista, burnished brass and copper boilers glistened about the room where soaring panels of Trompe l'oeil featuring pictures of Budapest's historic past, real or not, captivated the customers with myriad details of battles against the Turk and of castles on towering cliffs over the Danube. Potted palms trees lay scattered about the room, making oases of tables covered in dazzling linen and gleaming cutlery, as above huge, white Damocles chandelier fans hung suspended over the guests, gently turning with the air - with an unmistakable hint of menace.

"Ah … Coffee thank you," our grateful hero replied.

The three waiters vaulted away and returned seemingly immediately with three cups of coffee, a selection of cakes and buns, and the days selection of papers in German, Magyar and the Budapest Yiddish Chronicle.

"Just a single coffee thanks, really."

"Choose."

"Yes, you must choose."

"It is customary for the customer to decide."

"Yes, the customer is always right!"

"Well," Albert began tentatively, "this one ..." picking the one in the centre.

"Mine!" Yelled the right waiter.

"No! That was mine, yours is on the left!" Snapped back the centre attendant.

"I'm standing on your right! It must be mine!"

"Mine was down first! You put it over mine, on the left!"

"But I'm right-handed! And I never cross my cups!"

"You just did! This one's mine!"

The third waiter pushed a scone onto Albert while the other two were arguing, and as Albert somewhat perplexed accepted it - all three waiters burst into argument.

"He took the scone! Now his mine!"

"That's serve jumping!" "You cheated!" cried waiters one and two indignantly. "You can't do that!"

The argument went on for several minutes as Einstein watched his food and coffee juggle from hand to hand with the adroitness of a circus act. Presently, the maitres d' waltzed over and demanded they try feeding the guests instead of teasing them, this, however, only resulted in a long harangue from the waiters over minimum wages and the cost of living in an East European country.

"But I just want a single coffee," interrupted Einstein.

"Yes, of course sir!" The Maitre d' flourished with a spoon. "And a coffee you shall have – but first I must sack one of my staff!"

In a flurry of tea towels the three waiters whirled about and retreated into the kitchen, briefly leaving the Maitre d' as victor of the field of battle, the conquest was Pyrrhic as they suddenly reappeared with their mother in tow, who happened to be the cook, and the wife of the head waiter.

"János!" came an ear-piercing scream. "János! What are you doing to my boys!?"

"Oh no ... not Maja"

A spectacular argument ensued, with dishes flying about the room, palm trees were picked up and thrown, the glittering white table clothes ripped off their tables and tossed into the air. It was one of these that caught with the gearing of one the chandelier fans, and brought the fan to a snarling halt and cracked the fan's chain, so the fan came thundering down upon Einstein's table as everyone dove for cover. Albert sat there, still holding his coffee in utter shock.

At which point a whistle blasted the room from the front door and everyone came to attention as a trimly dressed Budapest constable walked into the restaurant, casually and persuasively swinging a baton.

"János! Maja!" he barked.

"Milos!" They all cried and then pointed at Einstein, and in vivid Magyar rapidly accused him of demolishing the restaurant and being in league with the Turks.

Milos, their cousin, walked over and poked Einstein in the shoulder.

"Passport."

Einstein dropped the cup.

In the cells, Einstein found himself in the company of a man wrapped in chains and a steel bucket over his head. This was the only interesting feature of the cell - the bucket was also the only toilet.

From underneath the bucket came a muffled voice.

"Can you take this bucket off my head?"

"Is it yours?" Albert asked.

"Err ... no, the guards stuck it on my head."

"Then no, I'm in trouble enough as it is."

There came a vigorous and pointless rattling of chains and banging of the bucket against the wall. He was later to be none other Harry Houdini the world's greatest escapologist, this was a future event, and at present he was still simply Erich Weiss failed escaped artist, which is Budapest is known more simply a jailbird.

"Take it off!"

"If the guards put it there! It must be for an official reason!"

"You have to be joking!"

"If I was kidding I would say something like, 'Two rabbis walk into a pie shop'.."

"Just take the damned bucket off my head! Damn you!"

"Well, if you're going to be rude about it." Einstein pouted and turned his back on Erich.

"Look! They stuck the bucket on my head to torment me! They're not supposed to stick buckets on peoples heads, it's illegal!" The bucket crowned escapologist argued. "And it stinks in here!"

Einstein turned around and yelled at the bucket. "If it's illegal, you should see a police officer about it!"

A heavy hand banged on the cell door, "Keep it quiet you scum, we're having lunch out here!"

From under the bucket muffled voice of the future Houdini continued. "Look, if you get this bucket it off my head, I'll help you escape!"

Einstein chuckled.

"No, really," persisted the young Houdini, "all I need it to get this bucket off my head and I'll be able to get these chains off, I've got a tiny key under my tongue. Once I've got that, I'll have these chains off in a jiffy, and once they are off, I'll get us both out of here in no time at all! Trust me, I know what I'm talking about."

Albert sized this up, and saw there might be a slim chance to it, after all things couldn't get much worse. He grabbed the helmet by the handle and began dragging it up over Houdini's nose.

"No! Arrrgh! Stop!"

"Now what?'

"It hurts – that's what! Use some lubricant, some oil or balm, I'm no good without a nose!"

A quick glance showed Albert there was nothing to be had except for a plate of cold Borscht, he scooped up a spoon full and attempted to spoon it into the bucket.

"Blu..rup! Stop! What are you doing?"

"Borscht."

A stream of Hungarian insults sounded under the bucket, until Einstein could stand it no more, and with a swift wrench popped the bucket off Houdini's head.

There was a thrilling moment of silence as they stared at each, and both recalled that fateful day in Florence, Italy when they had first meet, where Einstein had destroyed Houdini's stage act and ended up floating down the Arno river tied to a giant wheel.

"You!" shrieked an astounded Houdini.

"You!" cried Einstein and stuck the bucket back on Houdini's head chortling with glee.

The magician struggled violently with the chains, screaming in Hungarian about what he would do to Einstein, when the door opened, and Milos now promoted to the Budapest head of police stuck his head in.

"He's not out yet, I see," he grinned at Houdini, then pointed at our hero, "come on you. You have to meet the magistrate."

Einstein tripped out the door, leaving the receding fury of Houdini rattling and screaming behind him.

The court was dark and grim, the few candles illuminating the faces of the judge and accused only strove to enhance the gloom. The judge himself was a shadow of a wig and cape, which floated above his papers wafting his gavel like a 'Punch and Judy' show in slow motion, his wheezing voice only enhanced the "Pulliciniello" features yet somehow demanding absolute respect. Glittering eyes peered into the gloom to size up the errant Einstein, who clasped the rail of the accused stand with knuckles so white they gleamed in the candlelight.

"The accursed, "the judge jotted down a note, "is found guilty of … what?"

The court stenographer stood up and whispered in the judge's ear.

"Ah, yes, "the judge continued, "the Accused, not the accursed, is found guilty … oh now what?"

More whispering.

"What do you mean we have to have a trial? I'm a judge aren't I?"

The stenographer held up a piece of paper in front of the judge's face and pointed out the actual state of the case.

"Ah, I see," began the judge again, "it seems the defendant is 'still' a defendant." He chuckled and glanced about the courtroom. "Where is he? Ah yes. Now then, well, Herr Einstein, …, Einstein? What sort of a name is Einstein, doesn't that mean Mr One-Stone? Ah yes, how do you plead?"

"May I know what the charge is, your honour?" Einstein asked worriedly.

"Don't you know?"

"No, your honour."

"Really, oh, someone tell me what the charge is?"

Milos the constable stood up and barked. "The accused is guilty of declaring war on the Glorious state of Hungary, damaging restaurant property, placing numerous Hungarians lives in jeopardy and not paying his luncheon bill."

"Ah, didn't pay his bill," scrawled the judge in his immense volume. "And declared war on Hungary. Yes. Now then what do you have to say to that? Hmmm? Hey, wait a moment, how can he have declared war on us? Is he the Kaiser? Is he Bismarck? Doesn't even look like Otto, I've met Otto you know, not someone you want to meet on an empty stomach."

"He invaded our Glorious Nation without a passport - your honor - riding a bicycle!"

"Really" came a droll reply, "a bicycle no less. You're an idiot aren't you constable?"

"I try my best, sir!"

"Yes, and your doing very well at it, I can tell you. Well then, how much was the damage to the restaurant? And what's the bill?"

"Not guilty! Your honour!" Einstein cried, realizing it was now or never. "I didn't do it!"

"Oh bother, that means a long trial. Oh well, alright, call the first witness."

It was discovered that the witnesses were too busying serving lattes and crème tarts to attend, so a proxy witness was called in to stand for them. There was only one to hand.

"Why is that man wearing a bucket on his head?" quizzed the judge as Houdini was brought in.

"He is the witness, your honor!" the head constable saluted.

"Yes, but why is there a bucket on his head? That's what I'm asking."

"Lifestyle choice, sir!"

"Did you put it on his head?"

"Yes sir!"

"And why, pray tell, did you stick a bucket on his head?"

"He wouldn't answer our questions, Sir!"

"Constable, I want you to arrest yourself and take that bucket off his head."

"Yes sir!"

There was a lengthy pause, where everyone stood still, and looked at the head constable uncertainly.

"Well? Now what?" quizzed the magistrate.

"Sir! I'm under arrest, I am unable to remove the aforesaid bucket from the witness' head, Sir!"

"Blast, the sheer cunning of the man," twinkled the magistrate, "very, well, I want you to release yourself, remove the bucket from his head and then re-arrest yourself for obstructing the course of justice. Can you do that?"

"Yes sir!"

"Very well, and not a moment too soon, now who is the witness?"

After some prompting, Erich Weiss gave a false name, having too many creditors to take a risk.

"I am the Great and Amazing Houdini! Your honour! Famed escapologist and cape waver!"

The judge peered down at Houdini, then audibly scribbled in his tome, sucked his teeth and looked closely at the witness.

"I see now why they stuck the bucket on your noggin. Now, being an escapologist and all, and this question sort of begs itself, doesn't it, why are you still in chains?"

"Sir!" Milos the Constable saluted.

"Aren't you under arrest?"

"Yes sir! However I have information that is vital to this trial! Sir!"

"Very, well you're now un-under arrest ... can I say that? ... never mind, what is this vital information?"

"He was arrested for disturbing the peace, and the boys and I put the chains on him to see if he could get out of them, sir!" the Constable beamed.

"That's it!? That's vital?"

"Yes sir!"

"Well, now we know, I suppose it will have to do," and turned once more to Houdini, "alright then, lets see you escape?"

"The Great and Amazing Houdini will not reveal his secrets! I must respect the Code! Handed down from generation to generation, the secrets of the.."

"Yes, yes, I see, so what you're really saying is your stuck aren't you?"

"Not at all! It was the bucket, that's not normally part of the act, I could do it now in a thrice, your honour!"

"Really? Oh go on, show us, I love a good thrill, all though at my age getting out of bed and walking to the lavatory seems to be all the thrills I need."

"If your honour and the court would turn around for a moment, I the Great and Amazing Houdini! Ninth Wonder of the World! The Human spectacle! The .."

"Oh, just make it snappy, I have a bladder the size of a pea you know. Everyone, turn around and face the wall. I know justice is supposed to be blind and all .. but really!"

The entire court turned around and held their breath to listen to what was happening behind them. A furious rattling of chains and stamping of feet, grunting and swearing continued for several minutes, until it was rewarded with a explosion of anger and was followed shortly after, by the sound swift feet dragging chains out of the courtroom and into the street, to gradually fade into the distance.

"He's buggered off, hasn't he?" The judge wryly opined to the wall, and turned around. "Very well, next witness."

"Your honour! We call the defendant to the stand!"

"Really, this is going nowhere. You can't call the defendant to testify against himself. Well, maybe you can, but it makes no sense at all. Unless of course, you have a better a suggestion?"

"No sir!"

"Oh well," shrugged the weary judge, "what does the defendant have to say for himself?"

There was a lengthy pause as Albert digested the possibilities and came up with the only solution he could think of.

"I didn't do it, your honour."

"Good enough for me, case dismissed." He rattled the gravel and drifted out from the court, in search of a convenient and strangely thrilling wash room. Einstein was goose-stepped to the front door of the court and given a bicycle licence for identification.

Descartes and the Demon

The Rhythm method doesn't work – R. Descartes

A low moan came from the bed, the Sun had risen too early and the morning was not welcome beneath the covers. The face of René Descartes poked out from the sheets. He had an architecturally interesting face, with the sort of lumps and extrusions that the artist Hieronymus Bosch might have found exciting; warts dotted the skin, not so many as to create repulsion but enough to generate comment from certain species of hog. The ears, though hidden by long and lank hair, stood out in much the same way as birds have wings. His nose, well that was the question - was it a nose? After all, there comes a point in the evolution of every species when it ceases to look like the parent species it came from, as humans do not look like frogs - so Descartes nose ... well, was it a nose? With a face like this, his only choice as a professional career was philosophy. The pulpit was out of the question, as too many ladies would faint at his sermons. He could never be a lawyer, as the jury would spend the entire trial staring at his nose. He never felt comfortable as a professional soldier, as his dragoons kept falling about laughing as he gave them orders. In the end he had decided on philosophy, first because it made him a man of letters, hidden from the general public, and secondly because it made him irresistibly sexy in the short-sighted eyes of certain literary minded women and thus ensured an endless supply of crumpet.

"Oooh."

It was a long drawn out sigh that could have signified anything, but it was immediately followed by the bed sheets being thrown to one side and him leaning out of his bed to throw up the contents of the evening's before - into a chamber pot. It didn't take a scholastic philosopher to draw a conclusion. He lay back in his bed and tried to stare at the ceiling. Wondering why his brain kept moving inside his skull even though he was lying perfectly motionless, he tried holding his head in one place, but his cranium kept bouncing about until he realized he was still drunk and leaned over to regurgitate some more.

It was several minutes later before he was able to make the short progress to the room bell rope and ring for his servant.

"What?" - the unshaven and uncaring face of Pierre prompted from the side of the door.

The man in the bed pointed at his head and groaned.

"Haircut?" asked Pierre.

A groan in the negative.

"Shave?"

A double negative groan.

"Chocolate?"

There ensued a gentle sigh of approval, and a complete abnegation of responsibility for the morning.

Descartes was not a heavy or habitual drinker, he rarely imbibed the bacchanalian spirits, and he almost never caroused the nights, singing lusty songs and falling over. He never got the chance, he was such a cheap drunk that the smell of a wine wagon trundling past his window was enough to make him swoon. An actual glass of wine would have him swinging from the thick wooden rafters that crisscrossed his ceiling, swinging on them like monkey bars. He had a theory that because his Pineal Gland was so much larger than everyone else's, it absorbed alcohol at a much greater rate. He had many theories, most of them irrational at some level, but they all put him at the centre of an indifferent universe. This however did not stop the prodigious output of his pineal gland which was a legend in it's own anatomy.

"Why? Oh why?" he said to no one in particular, this made

him feel cleverer than he normally felt and he continued the internal monologue with his pineal gland. "How did I ever come to this?"

"You went out again with Monsieur de Spinoza last night."

René stared the ceiling in amazement; his pineal gland had never spoken back before.

"But it was the anniversary of the capture of ..." then he looked at the door and saw that really it was Pierre, returned with his morning chocolate drink.

René pointed at the table beside his bed.

"We will be having another day in bed, Monsieur?" asked Pierre with unashamed sarcasm.

"Ah, yes," René moaned and followed with a lugubrious sigh. "If only I could figure out God's great plan for the World. Ah, ... ah, the money I could make."

Pierre sat on the end of the bed and pursed his lips as he blew on, then sipped Descartes' chocolate. "Perhaps, master, the answer lies in our understanding of the natural world. Seems to me, that Monsieur God would leave clues to understanding his great plan all around us. We only have to look..."

"Hey, that's my chocolate," pouted René.

"Ah no, Monsieur, you decided against it and offered it to me."

"I did?"

"Oui, Monsieur, just a moment ago before you went philosophical again."

"Blast, well I've changed my mine again, go get me another one."

"Of course, Monsieur," Pierre remained sitting on the end of the bed and casually asked, "You are spending the month in bed, or only the week?"

"Haven't I been out?"

"Only to the brothels to go drinking with master Spinoza."

"Oh right, well it's hard work all this philosophising, you know, have to blow off steam somehow. We'd go mad otherwise."

This was met with a cheerful grin from Pierre, "Of course, master Spinoza does not drink and is a happily unmarried philosopher."

"Yes, well..." Descartes began philosophising again. "Tell me Pierre, how do you know you exist? I mean, dash it all. I'm in a bed, it could just all be a dream."

"Flatulence, Monsieur."

"Wasn't me, besides I'm under the covers - how could you smell?"

"No, Monsieur, I meant in regards to myself as a measure of existence."

"Well," Descartes continued, not at all convinced or even understanding, "open a window, you're the manservant. It's part of your job description and what-not."

"No, Monsieur, what I meant is that I fart, and by letting off steam, so to speak, thus I am reassured that this is not a dream. For I believe that flatulence is such an everyday ordinary thing that it would not be the subject of a dream. Or at least not mine."

"I see, so you fart - therefore you art," René frowned and pondered it's meaning. "I say that's rather good. I must write that one down. No, on second thoughts, you write it down, I've got this god awful headache."

"Oui, Monsieur."

The day morning slipped by the window, making no change to the day other than pointing the sun in a different direction. Eventually René sat up in bed and picked up a book. This he threw to one side followed by another and in turn a dozen more until he came on a book by François Rabelais. He quickly thumbed to the more bawdy songs and spent precious moments laughing.

"Monsieur is awake – again?" came Pierre again at the door jam.

"Yes, yes, today there shall be much philosophising, I predict," grinned René, "tell me Jan, what do you think I should think about today?"

"It is Pierre actually, but never mind," he sat on the edge of the bed with a chicken wing in one hand a glass of Bordeaux in the other. "Hmm, lets see, yesterday you solved the problem of the immortal soul by hypothesizing the pineal gland is the only connection between the soul and the body, hence the soul must be made of flesh, and thus not immortal."

"Oh I'm brilliant," René clapped his hands.

"The day before you proved the existence of God, because God created souls. Since souls are connected to the pineal gland and since you have a pineal gland, ergo God must exist."

Pierre dipped the chicken wing in his glass and chewed on it.

"Yes, yes and did you write all this down?" René asked with sparkling enthusiasm.

"Oui, Monsieur, as you instructed," Pierre continued with a mouthful of chicken and wine, "and the day before the day before, you demonstrated from first principles the transubstantiation of the pineal gland proved the primacy of Man over Baboon."

"Ah Jacque, we live in such ponderous time."

"Oui, Monsieur, and I am still called Pierre." There was a pause interrupted by a sip of wine and the wry question. "Monsieur, is it not possible you are putting too much emphasis on the pineal gland, especially since you have never seen your own pineal gland? And you have never actually shown why the pineal gland is so important?"

"Why? Why? You ask why?"

"Oui, Monsieur, I believe I just did."

"Because it is the seat of the soul, of course, it is an elementary proposition as adduced by careful reasoning, how much proof do you need?"

"At the present moment in time, any proof at all, would suffice. An assertion is not a proof, it is merely a statement."

"Bah, go make me breakfast, I'll have an answer ... no wait, I'm the Monsieur, you just go and make breakfast. I have thinking to do."

"Of course, Monsieur. Chicken sautéed in Crème Blanc with mushrooms again?"

"Oh rather, and don't skimp on the Crème."

Descartes was left once more to ponder on the seat of the soul. He had been doing this off and on for the last thirty years and had come to look upon himself as almost a guru on the subject. Very early on he had discovered that lying about in bed required the least physical effort, which in turn meant he could focus all his energies to discovering God's great plan for the world. Even walking to the privy was an unnecessary expenditure of his great wit, and he had reduced all bowel movements to the chamber pot, this vastly increased his philosophical output to the point where he had decided that no metaphysical conundrum was beyond his extraordinary abilities – even if he had no idea what the word 'conundrum' was supposed to mean. The outpourings of his vast and fertile intellect were relayed to an astonished world audience by his trusty factotum Pierre. René also had no idea what a factotum was, but he did know that Pierre made wonderfully, delicious hot chocolate.

On cold days Descartes would hide beneath the bed quilt with only his enormous nose poking out as he ruminated over the grand questions. On very cold days he retreated to the interior of a giant stove which sat in the corner of his bedroom, which was so warm inside it was almost a sauna. On freezing days he set fire to the house and moved the next morning.

In some respects it was Descartes who created modern philosophy, if modern philosophy is reduced to lying around talking nonsense while drinking chocolate. His main problem was, of course, that he had completely eliminated the real world from his vast and uncharted mind. As a consequence of which, he often had startling novel insights into God's great plan as he like to describe it, often startling divergent from God's actual great plan. It was along these lines, that staring into his chocolate cup one morning, he conceived his theory of Vortices. He noted by stirring the foam in his mug that bubbles swirled around surface in a manner not unlike planets orbiting the Earth. This was enough to convince him that invisible bubbles drove the planets. It's amazing what you can discover with an idle life, a hot cup of chocolate and smoking a tobacco pipe.

He stared at the window and wondered if he could close it with a stick. He stared at his walking stick and wondered if he could drag it towards himself with a pillow. He fluffed his pillow and fell asleep again.

At lunch, he was woken to the smell of breakfast.

"Breakfast? Already?" he said, "Bother, I had hoped to solve the problem of celestial movement by now."

"My apologies, Monsieur, shall I come back when your bowels are in the right orbit?"

"Best not," Descartes shot back, "I might swoon from starvation. I say, that smells good. Pheasant?"

"If pheasants swim in the Atlantic Ocean, are caught by fishermen in boats and are as flat as one of your books. Then pheasant it is."

"Splendid!" Descartes clapped, "quickly now, I am feeling cleverer this morning than I have felt for several months ... or is it hours?"

"Oui, Monsieur."

As Rene investigated his food he commanded Pierre to read the days correspondence.

"A letter from Cardinal Jean Richelieu, requesting you return one of his nuns."

"I thought we posted her back?"

"Perhaps the parcel has not arrived yet, Monsieur."

"Quite."

"Monsieur Galileo Galilei writes telling of his house arrest, punishment and torture by the Roman Inquisition and requests you intervene on his behalf with our king, to smuggle him across the border to freedom in France."

"That clever little bugger from Italy?"

"Oui, Monsieur."

"Then the answer is no. Got to keep the competition down, you know.

"Queen Christina Maria Alexandra of Stockholm, regent of all Sweden, Denmark, Finland and Norway of the House of Vasa, sends her warm regards and invites you to Stockholm for the privilege of personally tutoring her in your philosophy."

"Oh, that randy little tart is still after me," Descartes mused over his fillet of fish, "still - might be a few bob in her. Pity her nose is even bigger than mine and I've got quite a snoozer."

"Oui Monsieur, your rhinoliths are most distinguished."

Descartes considered his manservant for a few moments as he struggled to decipher his meaning, "...ooh my nose. Right."

"And a letter from our landlord demanding we pay the back

rent for the last six months or face debtors prison."

"In that case, we're off to Sweden."

"Oui, Monsieur. I have packed already."

"Excellent. How do we get the bed out the window?"

"We could set fire to the house again, and have the fire brigade carry you out-like last time."

"Hmm. I seem to remember I almost burned to death that time. No, we need an even clever plan than that one."

"We could try turning the bed sideways and sliding it out the door?"

"Won't I fall out?"

"Not if we tie you down?"

"Ooh, that could be fun. Run down the post office and see if they have posted off that nun. She was good with knots, I seem to recall."

"There is the matter of writing today's chapter."

"Which is?"

"Don't you know, Monsieur?"

"Aren't you my secretary?"

"Possibly."

"Close enough. Now tell me, what are I writing about today?"

"Very well. On the Discourse on the Method of Rightly Conducting the Reason in the Search for Truth in the Sciences, chapter seven."

"Hmm, bit of a mouthful, can't we cut it down?"

"Discourse on the Method?"

"No, it won't sell."

"The Method."

"Oooh, I like that," Descartes grinned, "Could mean anything and yet it has a definite ring about it. Stands out on the bookshelf. Get a few bob off that one for sure. Now read back what I last wrote."

'… thus we have shown the Pineal Gland is the seat of the soul.'

"I've forgotten, what does that mean again?"

"God knows."

"Clever. Yes, can't argue with God. Very clever. Even I'm flummoxed."

"Oui, Monsieur."

"But we have to throw in some mathematical gobbledygook just to make sure no one understands what I'm on about. Otherwise there's no lecture tour and that's where the crumpet is, let me tell you. Speaking of which, is there any crumpet?"

"There was, Monsieur, it was delicious."

"What – you had it?"

"Oui, Monsieur, as per your instructions."

"Are you sure I told you to eat it?"

"Oui Monsieur, almost as soon as you woke up, requesting I clean out the cupboard."

"Bother, I can't remember and you could be telling me anything. Well, where was I?"

"Mathematical Gobbledygook, Monsieur."

"Yes, mathematics is great for confusing people. Makes one seem much more cleverer than one is, not that I'm not clever, after all … oh tosh, what was I talking about again?"

"Mathematics and crumpet."

"Ah, crumpet."

"Perhaps Monsieur could introduce a completely extraneous idea with a mathematical framework, as opposed to discussing what other authors have gone over."

"You mean like crumpet and geometry?"

"Hmm, well, you are looking for a window into the soul,

perhaps windows?"

"Windows? What have windows... oh I see... windows could mean anything... that way our audience will fill in the gaps for themselves. Clever, glad I thought of it."

"Oui Monsieur."

"So what does one say about windows? Hmm, then what, no wait, lets throw in what you said before about farting."

"Perhaps Monsieur might substitute something more abstract for flatulence."

"I can't think of anything. No wait, how about Thinking? Can't get more abstract than that. I fart therefore I think? No, I think therefore I fart... no, no,... hmm, hard one this."

"Substitute, Monsieur."

"Oh right, I think therefore I am. Gosh I'm clever and it has such a ring to it. Got to have a ring, that's where the crumpet is, you know."

"Oh course, Monsieur."

"Lets see, 'I' ... that's me, 'Think' ... what I do ... 'Therefore' ... a word meaning philosophical gobbledygook which no one can argue with ... 'I am.' Which brings me back to me. It's brilliant. It says nothing, not only has a ring but is circular, and can mean bloody well anything I say it means. Oh, the crumpet this will bring."

"I."

"Go on."

"I meant 'I', as is me, perhaps you should define 'I' more clearly, Monsieur. It could lead to some misinterpretation."

"Zeus in heaven, no! You don't want people to understand! That's the last thing you need. No crumpet in that at all. Philosophy is like Religion – only without all that annoying incense and buggering the choirmaster. What's Religion? Hold up two sticks, tie them together and tell people it signifies God. The serfs fall for it every time. Make it too difficult and they start to question you. Oh no, no, no. Simplicity and circular reasoning that's the way to play this game. As soon as you appeal to the external world then the crumpet leaves by the nearest door."

"But Monsieur, I feel the logic may be too flawed to even work at the simplest level."

"Eh? Have you been reading books again?"

"You say you exist because you think."

"I was inspired."

"But what I mean, Monsieur, is that you could be dreaming you are thinking."

"My head hurts."

"But you might agree that a dream exists less than the real world, and a dream within a dream is even less of an existence..."

"Now it's a splitting headache, no wait, it's just a hangover."

"So, if we follow this logical deductively, we can never be sure how many levels of dreams we are from reality where existence is assured. In fact, we are infinitely uncertain as to the very nature of existence itself."

"There's no crumpet in this, now go write up what I said. We're off to Sweden in the morning. Is this bed on wheels?"

Newton and the Case of the Student Prince

Magdalene College is wheere Hell begins – I. Newton

The hours slipped from the clock with the swiftness of molasses on a cold day. Newton the student looked out of his window at the snow and wondered if classes would be cancelled again. For several days now he had been attending lectures, only to find he had been late for all of them, and he was a little worried the university had either moved to Wales or the Bubonic plague was back in town. Newton, as a young man, was shy and retiring almost to the point of being invisible. He was so diffident and unassuming that people often had trouble making him out from a cupboard. He also had none of the anger that seemed to radiate out from his glowing white hair of his later life, but that was largely due to a surreal amount of mercury poisoning from his alchemical studies. He was also shocking shy of women, so much so that he had never met one.

His room mate and the closest he had to a confidante, was lying languidly on the couch, smoking a pipe and sipping sherry. John Wickins was a man who knew how to live and he did this by living off other people.

Newton looked despondently out the window and thought he could never make the hours pass. "Wickins," he said abstractly.

"Golly, yes."

"Why golly yes?"

"Because you about to tell me something tosh interesting." Wickins looked bored. "I thought I could cover a lot of ground by giving you an immediate answer."

"Hmm."

"Rather."

"Do you think classes will be put off again today, with all this weather and all?"

Wickins looked up sharply at Newton with new fascination. It was like discovering a weasel in your morning porridge.

"Newton, it's the middle of winter." Wickins grinned. "We haven't had classes for about a month. We call these midterm holy days, and if my family weren't so bally bothersome I'd be home now doing exactly what I'm doing now. Which I should point out, is keeping you company and staying drunk."

"Hmm."

"Rather."

Newton pursed his lips.

"So yesterday, when I was in class?"

"Weren't."

"I was wondering where everyone else was."

Wickins grinned and poured himself another drink. The wine was Newton's, who still being a teenager had yet to understand the true meaning of theft. Wickins, in turn, was a perpetual undergraduate and parasite, and having discovered how absent-minded Newton was, had moved in the next day, and become his best, only and most expensive friend.

"Really Newton, you'll make a superb lecturer one day, I only hope I'm still a student, so I can sleep through your lectures."

"Hmm." Newton said dryly. "I expect you still will be."

"Touché."

Newton thought for a few minutes.

"What should we do?"

"We could have a picnic on the punts?"

"Isn't the Cam frozen?"

"Bother."

Newton started to look very worried.

"So classes are really off?" his voice almost rose in panic.

"Tosh, Newton, you really must have to get a social life that doesn't involve algebra, sinus, and mathematical philosophising."

"What else is there?"

"Well, trollops and drink of course."

"Trollops?"

"Yes, you have heard of them, haven't you?"

"Only in the abstract."

"To the pub then."

"Do they have a blackboard?"

There were many pubs and tavern within Cambridge, there used to be many more, but the crown ruled that using most of the corn for ale was causing a famine and had shut down half of them to save the remaining population. This tavern was old, so old you might have expected to see Anglo-Saxon's complaining about Norman upstarts and reminiscing about the good old days under the Romans. Its thick oaken beams hung so low that a nimble badger might have trouble digging its way through to the bar. Worst of all, was the appalling layer of smoke from the dozens of pipe smokers that had gone beyond being quaint, charming and rustic to industrial smog. It was a time when Cambridge dons liked nothing more than getting sloshed and ending up in the gutter after a long night's whoring: things haven't changed.

"Have you never been to a pub before?" Wickins smirked, as he hunched over at the bar to avoid the low ceiling.

"Where are the lecturers?" Newton seemed out of place and lingered by the door.

Wickins laughed out loud and lifted a mug of ale.

"This isn't a lecture room, unless you're here to be lecherous."

To this the whole tavern burst out loud, and belted their tables with their pewter mugs.

"I don't understand," said Newton, his naiveté was marvellous.

This however was not lost on the tavern's wench. Hilda, who had a thing for the college boys and their dons, lingered by him.

"You soon will, old man," grinned Wickins.

"Ale milord?" Hilda drew her fingers across Newton's chest, and immediately he jumped back out the door.

"Nothing ails me, I'm fine!" and was about to scamper down the cobbled road and back across the river to the college, if Wickins hadn't run out with Hilda to grab him by the arm and drag him inside.

"Its alright old chap," Wickins laughed, "she hasn't the pox, well not yet."

"Hey," Hilda slapped Wickins, "none of that, I've me reputation to protect."

"Dear Hilda, a regiment of the king's guard couldn't protect your reputation." He propped up Newton in a stall, poured him an ale and sat back to watch the entertainment. "Now Newton, this is a female of our species. You should consider yourself very fortunate."

Newton was terrified.

"Can she do algebra?"

"In ways you can't imagine?"

"I can imagine an awful lot, mostly the coming apocalypse."

Hilda crept closer to Newton to the point he could he feel her heart beating through a bosom that could have deflected cannon balls.

"Milord, have you a lady?" asked Hilda, doe like eyes that would have done justice to dead trout.

"My mistress is Science."

"Whose she?" She poked him in the chest, "A wench at the

Aylsebury Duck?"

Wickins laughed at this.

"Hardly, Science is the realm of the mind, the endeavour of the wise, the highest aspect of human reason."

"A whore then?" Hilda nodded wisely. "She must be experienced."

Newton could only look confused, and although he hardly understood what she was talking about, and Hilda in turn had no idea what world he was on, they seemed a perfect couple.

"Newton!" Wickins interrupted gleefully and began singing, "You must drink! Drink! Drink to eyes that are shinning like a sows ear!"

Hilda slapped him, but given the usual compliments that were thrown at her, didn't seem to take it too bad.

"Now then mister Newton, ow' about you and me goin' up stairs, and look at me trousseau?"

Newton looked at the ceiling in amazement.

"This place has a second floor? I never thought hovels had second floors."

"Er'e now, don't you be like mister Wickins, he's a bad un' that one."

"Yes," agreed Newton, "I'm starting to reach that opinion as well."

"Now then, you best be comin' up to me room an' seeing me trousseau."

Newton would have run away at this point, but Wickins was holding tightly by the arm and enjoying himself hugely.

"Why by Great Zeus," Newton trembled, "would I want to see your wedding dress?"

"You get to see her changing into it!" triumphed Wickins.

"And out of it!" Hooted another man who collapsed in the

pew with them and burped. Samuel Pepys was a fat man with large jowls, the sort who spent their hours in idle gossip and gambling, and this was exactly what he was engaged in. "Wickins!"

"Pepys you dog, this is my good friend and financier, Isaac Newton."

Newton nodded carefully and wondered how he ascended to the level of monetary advisor.

"Allow me to introduce myself, I am neither dog nor whore, and in this tavern that can be both," smiled Pepys as he stood up and bowed, then ducked as Hilda threw a half pound pewter mug at his head. "Oh Hilda, I didn't mean you, that is an assertion for Master Newton to derive all by himself."

Another mug flew past his ear.

"Newton, I have heard of you," Pepys leant back in his pew, "you're one of those mathematical lightening calculators, are you not?"

"My degree is in mathematical philosophy, mostly Aristotle, Euclid, but I also indulge ..."

"What is the mathematical reasoning on throwing two sixes from twelve die?" Pepys interrupted.

"Who's dying?" asked Hilda, her hands trying to figure out where Newton kept his purse and Newton felt his skin crawling at her touch.

"No ones dying," Pepys pulled out some dice and rolled them on the table. "These die, tell me Master Newton, how do I know the best numbers to fall if I throw these on the table."

Isaac for his part was torn between a need to put as much distance between him and a puddle of rags called Hilda and this odd question of reasoning.

"One minus five divided by six to the power of the number of times the die is throw," Newton said without thinking and tried to pull his hand away from Hilda bosom where she had somehow lodge it.

Pepys and Wickins grinned at each other.

"So which is more likely, two sixes or five threes?"

"Two sixes," Newton said instantly and groaned at Hilda's affection, "can we go now?"

"Not yet," Wickins grinned madly and poured him more ale, "what is the likelihood of two dices adding to five?"

"One in nine, no really I think I left a book in the library."

Wickins and Pepys, however, had suddenly acquired telepathy as they both realized what a goldmine was sitting next to them.

"All we need is ..." began Pepys.

"And a mark ..." agreed Wickins.

"Whose doesn't know ..." echoed Pepys.

"Perfect," finished Wickins.

Newton and Hilda watched this in some puzzlement, but before they could say anything, Wickins and Pepys had waltzed across the room and grabbed the other drinkers in the tavern and brought them back.

"What are we doing?" asked Newton.

"It's a exam of your calculating abilities," said Pepys.

"Ooh, an exam."

"Gentlemen, call your dice!"

An assortment of drayers, skinners and bargemen clustered around their table, holding clay pipes and acute inebriation as a standard. Thereupon each of the alehouse drinkers would lay odds on the dice, Newton would give the odds and Pepys and Wickins doubled their profits. This went on for an hour before their marks realized what was going on.

"Witches!" cried one of the bargemen, who had it in for

anyone from Cambridge.

"What?"

"Err?"

"The door!" exclaimed Wickins and headed for the dawn; they dragged Newton out with them and disappeared into the fog with a posse comitatus in hot pursuit.

Newton spent the day dragged about by Pepys and Wickins from tavern brothel and back to tavern, with Hilda trying to slip her hands into his wallet, until the moon came out and Pepys convinced them to steal a wagon and ride back to college. The snow fell about in droves as they forced Newton in the cart.

"Wickins!" hissed Newton, as he was forced in the back with Hilda, "where are we going? And is this your wagon?"

"Yes to the first, and no to the second."

"Sorry?"

"We're going back to my digs," Pepys cheerfully whistled jiggling the bag of coins on his belt. "I know a certain don, who would just love to meet you Master Newton."

"I'd be delighted," Newton seeing a way to escape the endless wriggling of Hilda against him pointed out that women were excluded from the colleges. Pepys, however, had been an undergraduate at Magdalene College and was used to the idea of buggering as a means of social advancement.

"Magdalene isn't quite like the other colleges, there are many frocks there, and almost too many I would hazard." He poked the snowflakes as they fell about them. "In some ways you might almost think it was a convent."

Magdalene College was one of the newer colleges and smallest of the colleges. It seemed to exist in that shady undercurrent between a full university college and a bargemen's doss-house. They ended up by the river, parking the cart and stolen ox by the quayside of the college. Red bricks loomed in the fog, and there was a danger they might lose their way in the dark when suddenly the stars came out and the moon shone like a lantern to illuminate everything. To Newton's distress it seemed mostly to illuminate Hilda's enormous bosom and hideous face.

"I think we will be just in time for din-din," said Pepys as he wandered into a privet hedge.

"Superb!" said Wickins falling over and only just missed falling in the river when the others caught him and they all staggered towards the college's formal hall. The snow lay so heavily on the ground they had to drag their feet through it and occasionally lost one of the company in the snowdrifts. As they staggered, they sung ruder parts of Good King Wenceslas, and burst through the front door of the formal hall before collapsing in a heap to the general applause of the staff and students.

A formal dinner at Magdalene College was completely unlike any of the other college's of Cambridge as was possible. It was more akin being transported to the wilder and more expansive moments of the Roman Empire. No part of any organ or orifice of an animal was wasted or ignored. Dishes like Swan's Tongue, Stoat's Eyes, Eels and Puffins Stew; Grilled Beavers' Tails with Black, Purple and Royal Mauve pudding was a statutory favourite. Pigs would be stuffed with pheasants, which in turn were filled with hedgehogs, which in turn were stuffed with nuts, figs and marzipan. All this food was guzzled down with gargantuan quantities of wine and ale.

Wickins stared in amazement at all the food and drink.

"How do they pay for all this? Surely not student fees?"

"'Tis' Whoring."

"You have whoring here too?"

"It is a whore house!" triumphed Pepys. "That's how we pay for it!"

"Newton," Wickins grabbed his arm and grinned with glee, "I'm changing colleges tomorrow."

"I think classes may have started again," said Newton, "I think I'll just head back to get my books."

"Sorry Wickins," interrupted Pepys, "but to enter Magdalene College you have to pass the test."

"Test?"

"You have to bed every wench in one night."

Wickins jumped with joy - as Newton fainted with fear.

Newton woke up in a room filled with young women and most of his clothes missing. They seemed to be dicing for his clothes. It was the worst moment of his life, a room filled with women, his clothes missing and not a textbook on Euclid in sight.

"It's Golgotha!" he screamed with messianic fervour.

The girls looked up, almost relieved they had something to do.

"Oh er'e now," said the one with the fewest warts, "young master has gone and woken up."

"He's mine!" said Hilda with an addled sense of possession, as she burst in the room, leapt across the room with his pants, and prepared to defend her customer.

"E're now," explained the Warthog, "that's nae how it works. We all get's our turn."

"Nae this one," Hilda picked up a candle and brandished it dangerously, "I got 'im first!"

"No need to be like tha'," said the Warthog as she pulled out a poker from the fireplace, "let's roll for it."

"Roll?" asked Newton.

Three hours later as the sun weakened in its battle with winter shone across the chimneys of Magdalene College, Newton walked out of the college holding several dresses and a purse filled with coin.

"There should be a law against this," he muttered angrily, and trudged his way through the snow back to Trinity. In so doing, he became the only undergraduate to pass through the halls of Magdalene College and not lose his virginity.

Einstein and the Patent Office

The speed of light is still the same
even when you hit yourself with a hammer – A. Einstein

It was a dark room, lit only by beams of light sliding from shuttered windows through a haze of dust. It smelt of paper, dried paper, new paper, wet paper, moldy paper, paper mâchè, endless reams of toilet essays, wads of manuscript, all made from cotton, hemp, linen, rice and occasionally ancient socks. Paper so old it had become translucent with age, almost to the point of evaporating from mould, giving off a pearly luminescent glow of decaying bacteria.

Einstein had walked into this gloom, holding a scrap of paper. He was there for a work interview.

Unable to return to Germany, as he couldn't find a way to get to the Berlin Office of Passports, Immigration and Library Fines without first having a passport, which was of course, impossible. He was forced to find employment in Switzerland, this being the only country willing to take in a stateless person.

After all, who else would conduct the trams?

"Hello?" He spoke timidly into the shadows.

"What? What? What?" Came an antique voice.

"I'm here about the tram driver's job?" Einstein said. "Is this the right place?"

A candle wafted across the gloom, which was suddenly besieged by a flock of suicidal moths.

"Bother," came the antique voice, "drat these moths."

There was a scratching sound, followed by the flare of a match; and a nose bomber moth with nothing to live for instantly extinguished this.

"Ach! That was my last match. Never mind, turn on the lights would you. It's that switch next to your elbow."

Einstein fuddled in the dark for a moment and then with a snap, the room was flooded with illumination. He blinked in astonishment at the abrupt dawn.

"Why don't you have them all the time?"

"Pigeons."

The fluttering of wings followed this comment, as several white pigeons descended from the rafters and began strutting across the endless reams of paper.

"They nest here, causes me no end of headache with their cooing and their fluttering. I'd buy an air rifle, really I would, but they're a protected species, you know."

"Pigeons, a protected species?"

"Don't ask me. I just work here." The wizened old man clerk spoke with no little annoyance. "Now then, come about a job, have you?"

"Yes," Albert held up his piece of paper, "you have advertised for the position of tram driver."

"Can you drive a tram?"

"No."

A scratching of paper followed this.

"So, why should we give you work as a tram driver?"

"I can ride a bicycle."

"Is that like a tram?"

"No."

More scratching.

"That's probably not the sort of qualifications, we are looking for. We really do need someone who can drive. Even just a little bit. Can you drive a tram even just a little bit?"

"Ah... no."

"Kinds of makes it a bit difficult, don't you think?" The ancient secretary pursed his lips, "Hmm ... I know we have a position in one of our offices checking patents for inventions and gadgets. You don't need a licence for that. Just show up and spend nine hours a day, stamping papers."

"I could do that!"

"Why doesn't that surprise me? It's on the other side of town, just catch a tram ... did I ask you if you could drive a tram? I did - didn't I? Bother. "

The doors to the Patent Office were a gift from a well furbished Scottish industrialist called Maxwell, who designed the large glass revolving doors to be driven by powerful hydraulics and electrical motors, this terrified anyone who was not used to their eccentricities. On most days Einstein would leave work almost as soon as he had arrived. Leaving him gasping in frustration, as he shot back out from the foyer staring once more into the snowy Bern street. He would turn around to find the building was still where it ejected him, and would charge back in, only to come flying out again at even higher velocity.

"Gott im Himmel!" He would cry at finding the street once more, which he had hoped to leave the previous moment. "Where is the demon running this thing?"

This carousel would resiliently prevent Einstein from entering the building, until by a random chance he would shoot straight through and fall sprawling into the main foyer. The doorman, whose sole function was to direct people to the elevator, always greeted this event with applause.

This elevator in turn was a gift from the same Scottish inventor, and was based on the new hydro-pneumatic lifting principle, where the weight of the lift and people riding it, was balanced by and equal weight of water and air pushing it. Due to a poor translation of the user manual, from Scottish brogue to Swiss-German patois, it had been constructed upside down, and introduced such a violent randomness into its mechanism, it would leave the user in a state of paroxysm in simply trying to get to their floor.

The cries of "Waaaah!" would echo up the lift well, as its passengers descended with a button popping speed, followed by "Aieeee!" as it shot straight back up again, before ejecting it's occupants out with an enthusiasm that bordered on cannon fire.

Albert would tenderly make his way to his office on the fourth floor. It was a tiny room, almost an alcove, hidden away from the world, with only a tiny window overlooking the streets of Bern. Outside this window, storks would nest in summer and poke their heads in the window as if to ask what he was doing there, and drifts of snow would fill it up in winter, but occasionally, weather permitting, he was able to look out over the red roofs of Bern, past the tall steeple of the Münster Cathedral at the distant mountains whose glittering peaks shone like marble in the sun.

Most mornings, he sipped his coffee, staring out his window at the bank on snow on the windowsill. Sighing in contentment, he would let the mornings work fly past, as the occasional gust of wind would scatter the papers about the office in wild abandon. Since no one ever bothered to check up on what he was doing, he was able to sit back and let the mornings ease into midday. Staring into his cup and letting the crème spin around the cup, he could make tiny galaxies of white foam spinning upon the black coffee.

His office was really a clearing house for the papers of the mad inventors, those inventions which refused to fit into the rigid system the patents office had set up to deal with everything from flying vacuum cleaners to disposable sewage farms. Every conceivable arrangement of matter and human whimsy might waltz across his desk in a day, much to his enjoyment. As it gave him no end of time to spend day dreaming, for should his supervisor ever bother to visit and say to him he must catch up on the paperwork, he could always reply he was thinking about the plausibility of a hydraulic motorized cork screw, and pick up a very paper on the subject.

One morning, as he reposed at his desk, a resounding knock came at his door and he fell over backwards out of his chair in surprise.

"Gott!" Came his inevitable response.

"Herr Einstein!" came a thunderous voice that carried tones of damnation and martial justice. Albert shot to his feet and for no reason he could think of saluted, and clicked his heels.

"Yes sir!"

"I am Herr Olsen, I am here about my patent."

A tall-distinguished man, graying at the temples and carrying a hat and stick, loomed in Einstein's door. He carried papers in his hands, obviously one of the standard patent applications that required at least a cursory glance.

"Ach, ja." Einstein hesitated, and wondered if the man had come to the right office. "Please, sit down."

There was only one chair in the office, and they both looked at it, for it was obviously Einstein's chair.

"But where will you sit?" Olsen the guest asked.

"I will stand," said Einstein with tremendous politeness.

"I couldn't let you. I insist, you sit, it is your chair." Herr Olsen persisted.

"Nonsense. You are my guest," and pushed the chair to Herr Olsen, who pushed it back with equal vigor.

"I could never deprive a man of his chair. It's uncivilized."

Olsen remonstrated with the chair.

Einstein, who never had guests in his office before, refused to demure. "I insist! You are my guest," attempting to push the chair back.

A skirmish began between the two as they pushed the chair back and forth with steadily increasing violence until one of the legs broke on the chair and they stood in a moment shamefaced silence.

"Coffee?" Einstein saved grace with a return to civilities.

"Ah, yes, please."

Herr Olsen stood rigidly to attention, as Einstein was able to bustle about the coffee stand, grinding coffee, packing the peculator and setting the coffee pot upon its burner.

"Now, Herr Olsen wasn't it, you have a patent for stamping," Einstein looked squeamishly at his guest.

"Yes, I am here to insist this be unstamped."

This was a new one, and it had Einstein stumped for a moment as he looked over his desk trying to find the stamp that unstamped, before he turned around and asked what this meant.

"It has been stamped saying it has been refused." Herr Olsen remained motionless as he spoke, "I have come from Copenhagen to tell you that this patent should be issued here, and there has been a clerical mistake."

"I see," Albert said carefully.

"This patent should have been issued. It must have the stamp which says it has been accepted!"

"Ach, mein gott, my apologies," Albert sighed and realized that one of his papers had come back at him. "Please," the coffee had begun to boil, "take a seat," and turned around to fix the coffee. Instead, Herr Olsen strode forward, pushed his paper under Einstein's face, and pointed out the signature.

"Here, see. You are Herr Einstein, are you not?"

Leaning back Einstein squinted at the papers, and there in the bottom corner in spidery script was his signature as plain as rowan tree standing starkly in the winter snow. He had a vague memory of it passing across his table, and an even vaguer memory of stamping it.

"Ah," passing the coffee cup to Olsen and taking the papers in return.

They were in fact a description for a 'steam powered land submarine', and it came back to him what an absurd idea it was.

"Ja," he said slowly, trying to find inspiration in the moment, "a submarine for ... going over land ... yes?"

"That is it precisely!" Herr Olsen stared down at Einstein with maniacs' eyes, and Einstein felt himself trapped.

"Can you ... err ... tell me how it ... umm. Floats?"

"It doesn't! It only goes on land! That's the whole point!" Olsen said with an unremitting intensity that bordered on religious zeal.

"I see," said Einstein doing his best to avoid the unblinking stare of the inventor, and slowly asked again, "So how is it a submarine?"

"What?"

"Well, 'sub'... means under ... and 'marine' means sea ... so, how is it an undersea boat?"

Olsen leaned closer, "Ah. I hadn't thought of that," then leaned back and stared at the ceiling, "Ah."

"Take your time." Einstein stared at his coffee as he stirred it about, "cake?"

The tall Dane about faced and strode out of the office in acute embarrassment, then returned a moment later and without a word gave the coffee cup to Einstein before once more marching out.

Albert let out a slow whistle and set about reattaching the leg to his chair.

A month later, his second inventor came a-visiting. Albert was sitting at his desk stirring his coffee and looking over an obscure paper on Brownian motion, when at his shoulder there was a sudden and all too quiet greeting.

"Hello," came a soft voice at his shoulder.

"Waagh!" Albert fell forward and then backwards out of his chair, to stare up at his sudden interloper.

"Hello?"

A balding man with pince-nez glasses looked down at him with an abstraction not unlike a naturalist discovering a new insect in the wilds of Patagonia. Albert, in turn, stood up, confronted the intruder with a more dignified response, and offered him a coffee.

"No."

"Tea? Herbal?"

"No, tea."

"Biscuit?"

"I'm here about my patent."

"Of course. Of course. Have you seen the clerk at the front desk?"

"No."

"Have you seen the clerk on the first floor?"

"No."

"Have you seen anyone else, before you came here?"

"No."

Albert sighed, as the man stared hypnotically at Einstein through owlish glasses - Albert was beginning to recognize that 'stare'.

"Your patent didn't get accepted, did it?"

"No," came the predictable soft answer.

"And you have come to convince me, to give it the proper stamp?"

"Yes."

"Well then, can I see the patent pending?"

There was a brief investigation of the man's tattered brief case, which produced an even more tattered collection of pale blue patent documents. Albert shuffled through them trying to find a recognizable diagram, or even a title; this perusal went on for several minutes. All the while, his interloping inventor gawked non-stop at Albert, which given his proximity Albert found disconcerting.

"I don't seem to be able to find, a cover sheet." Albert was forced to concede. "Can you remind me what it's about?"

"Don't you know?" Came the myopic stare.

"You could tell me, that way we'd both know."

"It's a new type of winch for a sailing boat."

"Interesting," Albert racked his head trying to remember what it was about.

"It halves the number of turns needed to wind up a winch."

"Very interesting, can you tell me when you submitted it, I could look up my notes."

"Today."

"What?"

"I'm submitting it today."

Albert sighed. "Right, take this downstairs to the front desk, and give it to them. I only handle the stamping, not the receiving."

"I know."

"And!!?"

"Can you stamp it now, anyway?"

"No. You have to go through the proper channels, the paper work has to be done, the coffee has to be made."

"Could you stamp it anyway?"

"No!" Then pushed him to the elevator randomly selected a floor and watched the inventor disappear, as the soft voice of his guest came over the rattle of the elevator.

"Fear me. For one day I will rule the world."

His third encounter with an inventor, was preceded by the clarion call of an automobile from the street, which in those days was still enough of a novelty for people to stop and stare as they passed by in the street.

Sitting at his desk in his office, remote in the Himalayas of the fourth floor, Einstein could hear the distant echoing of what was presumably a large man haranguing the concierge at the front door. This was soon follow by a first floor encounter with the chief clerk who relayed the automobile driver to his second clerk on the third floor, at which point a thunderous yawp could be heard rattling the cage of the lift and quite possible rattling the third clerk who very nervously could be heard saying the name of Einstein, who in turn on the fourth floor revolved in his seat and gripped the armrest to confront the now clear and present danger that was irrefutably bounding up the stairs.

"I'm here about my patent!!" Came the bellow of a red headed man, who with all the enthusiasm of a plague of Vikings storming into a monastery convinced they are about to plunder Byzantine gold, stomped into the room and leaned over the rapidly reclining Einstein.

"Have you an appointment?" Einstein attempted.

"I do now!" Came a manic gleam.

"Yes," Einstein sighed, "well then, take a seat."

The stranger did, and pulled Einstein out of his and sat in it.

"Thank you," said the interloper. "You're the first polite person I've met here."

"I'll just go ..."

"No wait, this won't take long, I need you ... you do authorize new patents, don't you? Good. Now I need you to look over my automobile. It's a new one, designed it myself, and I want you to give me the go-ahead, and sign a patent giving me all rights European, American and worldwide, from Istanbul to Tasmania. Are you following all this? Good, it won't take long, we can take a ride up to the mountains, get a breath of fresh air, you can look over the engine, and Otto's your uncle, I'll fly you back. Take two hours, won't even notice you've left the office. What do you say? - And it had better be yes."

"I'll just ..."

"Right, lets be off." The stranger grabbed Einstein by the arm, threw his jacket over his shoulder and waltzed him down the stairs, enthusiastically telling him of the time he had been on safari in Kenya and had accidentally shot his own game warden. In no time, Einstein found himself ensconced in a machine that seemed to be made from rivets, rods and vulcanized rubber, then driven at high velocity through the winding streets of Bern to the Alps, as concerned burghers dove for doorways in the narrow streets.

It was over an hour before Einstein was even able to discover his abductors name. The car screeched to a halt over an imminent precipice and the owner, one Mister Kraus Pots from Baden-Baden, pointed at the horizon and burst into song. Given it was an unexpurgated rendition of Sailor Sue, it seem totally out of place in the Alps with their glistening peaks and dreamlike panoramas.

"Kraus is my name! So, what do you think?"

"A little bawdy maybe ... do you mind if I throw up, I think I'm a little carsick?"

"Not the song! The car! The beast! The Machine of Zeus!"

The high speed, the high altitude and the hysterics, all caught up with the wayward patent clerk as he leaned over the side of the car and removed any remnants of his mid-morning coffee break.

"There's no time for that!" Pots dragged Einstein back into his seat. "The car! What do you think?"

"Well, fast, yes, it does seem to lack a car body, you know that bit that goes on the outside of a car. Other than that, you do seem to have a fast car."

"More than a car!"

Einstein looked about him, trying to discern what novelty had been incorporated into the dashboard.

"So? Solid pneumatic tires?"

"Much better! Try again!"

"Washable seats?"

"Not even close! One more, come on, you're almost there!"

Einstein sighed, "Well, oh I don't know, does it fly?"

"Yes! I knew you were smart the instant I laid eyes on your square head!"

"It flies?" There was a speculative moment, as Einstein considered this nonsensical idea. "You're telling me, this thing flies?"

"That's the spirit! It's my 'soon to be patented heavier than air flying machine'!"

Einstein leaned over the side of the car looking for wings, retractable or otherwise.

"Really?"

"Yes! And you're to be the first passenger of the world's first flying car! And you're to be the world's first co-pilot!"

This brought more alarm to Albert than the entire morning's drive.

"Ah."

"Hang on to your seat! This is it!!

The automobile revved its 30 horsepower 1902 Daimler-Benz

engine and shot off towards a cliff with an eagerness that would have impressed a greyhound suffering from arthritis. At the last moment, Einstein leapt from his side and threw himself onto the top of a pine tree, which flexed alarming under his weight before bowing out over the precipice with Einstein enthusiastically hanging for all his worth. As the car gracefully plummeted towards an alpine lake, Mister Pots called out indignantly:

"You fool! You spoiled the airflow!!" before shortly vanishing into a crystal clear Alpine lake.

He soon bobbed up again and shook his fist menacingly at the hanging Einstein, who was far too concerned with the imminent defoliation of his pine tree as he slid further and further towards its tip. It was only with an unexpected wellspring of agility, that would have impressed the Flying Zucchini Brothers, was Einstein able to sway back to the cliff face, and claw himself once more to safety, all the while wondering how being a patent clerk could be so dangerous.

Da Vinci Dum-Dums

Leonardo da Vinci wasn't just clever
no, he was insanely clever! - L. Da Vinci

His home was a workshop overlooking a thousand year old piazza in Florence, endless contraptions hung, sat, or were even wired to the walls, some of them so large they were the walls, held together with paper, plywood, papier-mâché and unlimited dabs of genius. A few of the devices fluttered or moved, driven by mice, or clockwork or the wind, waterfalls or possibly even magic. All of them ornately crafted or painted in delicate colours as phantasmagorical beasts and animals, a lion that bent down and opened its mouth as an egg rolled out, a cockatrice that flew up to the ceiling and back down again whistling and dancing. The tables were littered and stained with vials, ointments, unguents and mineral salts, giving off an overwhelming odour of mystical discovery. Screws, gears and odd completely unexplainable metal objects spilled from draws and cupboards with the exuberance of motorized pixies trying to escape the magical forest.

It was quite simply the sort of room every boy wishes he had, and the sort of room every mother wishes her son would clean up.

Leo was clever, clever in the way pigeons cannot only fly but fly head first into glass windows. Clever in the way hedgehogs cover themselves in an impenetrable array of spikes and then play chicken on a highway. Clever in the sense he thought outside the box, he thought so far outside the box some of his thoughts were off the paper altogether. He was so busy thinking ahead of his time, he never really came to grips with his own.

This is not to say, he was incapable of tremendous discoveries, or that his insights weren't visions far beyond the realm of ordinary mortals. The trouble was some of his ideas were so incredible and revolutionary, that none of them could have been constructed for another five centuries and beyond.

Some of his inventions were indeed unbelievable, the first bicycle, the first winged flight as in a school boy's paper airplane, machine guns - well, a bunch of muskets on top of each other -, an armored tank, parachute, cluster bombs, and a bridge across the Istanbul straits. He virtually invented the science of anatomical drawings. Some of his other contraptions, however, were not so spectacular, including the transatlantic milk-float using donkeys, his helicopter driven turnip, the discovery that lepers aren't what they used to be, and the first medical dissection of a shoe.

His real problem was his ego, his limitless sense of his own amazingness. He looked at himself in a mirror he specially designed and crafted himself, and flexed his muscles, marvelling at his own physique.

"Oh Leo, Leo," he said to himself in appreciation.

There was no question he was a handsome man, with his long flowing hair, piercing gaze and chiselled face he was eagerly sort after by many of the girls of Florence. Unfortunately, he spent a great deal of his time either inventing new forms of Art, reinventing new types of mechanics or as he was now, simply admiring his incredible self. "The world is so lucky, I'm here!"

Leo stared out through the Venetian blinds of his window onto the piazza of the Holy Annunciation. He wondered about connecting all the towers of the Florence with wire cables so people could send messages to each, then had an idea about directing the Arno river into branching canals to irrigate and provide transport for the farms then changed his mind and wrote a treatise on fortifications for Paris.

Three hours and a one hundred and fifty pages later he got up and stretched, and then scratched his name in indigo on the title page, 'I, Leonardo', he felt other no explanation was necessary. Merely a statement of his existence would suffice future generations in the same way the pyramids said 'here, pyramids, amazing eh?'

He admired himself again in the mirror as he nibbled on cheese. Then he drank a quart of goat's milk and sour cream, which along with some nuts and occasional salad was his entire diet. He devoutly believed in vegetarianism, and as a result suffered terribly from constipation and colic.

A knock came at his door.

"Not now, I really am a genius at work!"

The knock came again.

"I could be working a cure for the plague you know? You don't want to miss an opportunity for me to save the world, do you?"

Yet once more came the knock, and he sighed and moved across the room with the sort of complete sense of focus and sinuous strength that only lions seem to have.

"Well?" he barked as he threw open the door, then swept his cap to the floor, "Ah, your highness, please enter my humble abode."

An ego equal to his own strode into the studio.

Cesare Borgia was not a man to be ignored, not unless you happened to live in a country on the other side of the world and were prepared to spend your entire life in a suit of armour with your back to a wall. Blackmail, assassinations and outright murder were only a few of Borgia's many tricks for maintaining power.

He, in turn, was the son of a Pope who was so corrupt he had eight mistresses, had tried to sell the papal crown on at least two occasions and had probably committed his first murder by the age of twelve. Borgia, the younger, had been brought within the inner circle of a society so far from the teachings of Christ he might as well been brought up by the devil himself.

"Ah, Leonardo," Borgia sighed breathily in a manner that said little but communicated reams. "I have missed you at court."

One of the few men in Florence who was as handsome or as tall as Leonardo, stood within the centre of the room as if he owned it, which in a sense he did, as he was Leonardo's paymaster.

"My lord," Leonardo smiled convincingly, "forgive me I have been working on a little project for your entertainment."

"Wonderful, tell me, oh do tell me."

"I do not wish to spoil the surprise, my lord."

"Spoil me! Spoil me!" said a man who had once taken a crossbow and shot to death a group of convicts locked in a courtyard - just for target practise. Cesare Borgia was quite simply a spoiled and very dangerous child. Anyone brought up in the debauchery of the late renaissance papal vices was going to be very, very spoiled, in the way apples left in pigs manure for five weeks go off, so Cesare Borgia was bad.

Leonardo knew his life was in the grasp of a man with so little regard for human life he had trouble distinguishing peasants from cattle; on the other hand Leo had such confidence in his own abilities to outwit even the highest and most dangerous lord of the land he acted with impunity.

"My lord, should I? Should I spoil the surprise? The astonishment will echo through the courts of Europe like the cry of a titan."

"Yes! Oh please, please.... Hey wait up. I'm Cesare Borgia."

"Your majesty let me show you, you will be amazed."

Cesare Borgia jumped up and down again, and then got frustrated, pulled a dagger and said, "you know my papa is the pope?"

Leo smiled broadly, knowing full well he could disarm the young psychopath with sweets.

"Come this way my lord," he said with a flourish holding up a bowl of candied fruit, "I had intended to surprise you for your birthday, but for the son of the Pope every day is a birthday."

"Ah, you are a clever man, you know I was joking," grinned Cesare as he put away the knife, started to choose a sweet but then took the whole bowl. As he was about to eat one he looked sharply at Leo and then at the sweet, and the possibility of the sweets being poisoned entered his mind. Not that Leo had any reason to assassinate him, but poisoning was such a routine aspect of Cesare's own life that he put down the bowl and pulled out the stiletto again. "Are you not having one?"

"Of course my lord, tell me which one should I try?"

"This one... no this ... one," said Cesare in a moment of indecision.

Leo picked up a large lemon jub-jub. "This one my lord?" and was about to pop it in his mouth when Cesare shrieked.

"I want that one!" and tore it out of Leo's hand.

"Excellent choice," Leo smiled and then pointed with his finger, "If I may point out there is a surprise inside, my lord."

Cesare, whose fear of poison could be measured on the Richter scale, gave it back "You eat it!" and once more pulled out a knife.

Leo calmly cracked the shell of the sweet and a tiny paper bird flew out and fluttered to the ground. "You can eat the sweet as well."

"Fantastic!" Cesare cried, opened up one himself and another brightly coloured bird-sweet fluttered to the floor. "Fantastic! Oh, Leo you are such a treasure."

"Yes," smiled the artist, "I like to think that as well."

They walked out to the balcony overlooking the piazza of the Holy Annunciation and sat on the wall, popping sweets and watching them flutter to the cobbles stones beneath. From the window they could look down a narrow street to the swelling breast of the 'Basilica di Santa Maria del Fiore' or as the locals called it 'the Duomo', that great maternal symbol of the cathedral of Florence that bulged out from the red tiles of the city like a giant jelly boob. Indeed most of the architecture of Italy was of two forms, either enormous bulging breasts like the Duomo or a phallus as in the appropriately angled leaning Tower of Pisa. In other countries the locals built edifices like the Egyptian pyramids based on mathematical ideals, or the Norman obsession with ruddy great defensive castles, but in Italy sex ruled architecture like a mistress with a whip.

"My lord," said Leo suddenly, as they warmed themselves in the sun, "I have an idea for a giant walking man, covered in iron and filled with cannons that would destroy all your enemies." He finished casually popping another sweet.

Cesare blinked and looked at his host.

"Would it work? I mean how might it walk?"

"By donkeys deep inside it, walking on treadmills which pushed giant wheels and gears." Leo became animated. "A man in the head of the giant could pull levers and switches to control its movement, and other men could operate small cannons on its torso both on the front and the back. Giant hands could reach down over castles walls and tear them to pieces, and its feet could step across the moats and rivers in the same way you and I go walking in the street. It will be called the Colossus of Florence and with it you can conquer all of Christendom, even stride through the seas to retake Constantinople and Jerusalem."

Cesare coughed on a sweet.

"You do think big, don't you, and how long did it take to conceive of this idea?"

"Just then, as I was looking at the cathedral. Look at it this way, my lord, is not an army already a giant? A giant that is scattered all over the countryside or the sea in ships. An army is a large number of men who all work together in the same way as a giant. I propose putting them all in the same body, it would be like a castle that stomps across the country, putting your enemies to flight and creating an empire for you."

"And you thought of all this, just now?"

"Yes, of course, I am Leonardo."

The Duke scratched his chin and looked unsure.

"What will you call this giant iron man?"

"It will be a Transformator."

"And donkeys will pull gears that move its legs?"

"As men will throw levers to throw the gears that stride the world!"

"I see, and can you build this?"

"I am Leonardo."

"Yes, but can you build this within the next two hundred years?"

"Have I ever let your lordship down?"

"No, but I am still waiting on the mechanical birdman that flies, and the giant mole machine that will allow my armies to tunnel under castle walls and pull them down, or for that matter the cannon that reloads itself and fires continually. It's not that you have let me down, it's just I haven't seen any results. Although, these are very nice sweets."

"Ah, but my lord you have not given me the monies I need to construct these marvellous devices. Am I to construct them out of air, will the clouds furnish me with craftsmen, does iron grow in the steam from rain?"

"You are Leonardo, you tell me," Cesare smiled, and when a man like Cesare Borgia smiles it pays well to notice. "Now then, where is my present?"

"Another?"

"I am Cesare."

"Ah, yes of course. Hmm, how about a ring that makes you invisible?"

"Can you make such a thing? This sounds like witchcraft."

"No, of course not, but that's not the trick, the trick is if your enemies truly believed you had such a toy, they would never sleep at night, or day for that matter, and would be too terrified to attack you."

"Well, that is a surprising idea, so there would no ring, we just let people think I have such a thing. No, I have a better idea, make me such a ring, that will be a nice surprise."

"But your lordship, that is impossible, no one could craft this."

"You are Leonardo."

"I ... yes," and for the very first time Leo felt unsure.

"Good," said Cesare, "and when you have done this, I shall pay for your giant iron man. Shall we say, next week?"

"Next week, surely two weeks?"

"Make it five days."

"Done!" Leo said fearfully, before the date fell to the next day; one did not haggle with the Borgias.

For two days Leo pulled his hair out and hid under his covers in a blue funk, wondering he had finally overestimated his abilities. He thought of fleeing the country, running away to Constantinople or god forbid the Netherlands – with its terrible cooking – they didn't call it 'the nether lands' for nothing. The possibility of getting his mother to write a note asking Leo to be excused crossed his mind several times, but he knew the prince would just send soldiers with another note to ignore all mothers. He looked longingly at his winged flying apparatus, made from pigs bladder, bees waxes and string and wished he could get the bloody thing to actually work. Eventually, inspiration came to him, as it always did, when he started to think about the psychology of Cesare.

Leonardo appeared the next day at the young princes court with a small box painted in mystic symbols and astrological signs.

"You've actually made it?" the young lord stared in puzzlement at the artist. "Already?"

"I am Leonardo."

Cesare coughed in amazement.

"Well, show me."

"Of course my lord, the ring is made from gold and rare silver. I have studied mystical charts from the east and the movements of the planet Mecurius, Santurnus and the hidden stars of Luna to plot the wheel of your future, thus allowing me to hide your movements from the stars. It is here in this box, let me show you."

Leo opened up the box to present a velvet lining with a beautiful ring, delicately made; he picked it up with his fingers and held it up to the young Borgia.

"Here it is, my lord."

"A ring of invisibility? Truly?"

"Yes, my lord, and only you may use it."

"Why?" asked Cesare suspiciously.

"That is the nature of the magic forged by alchemy within the metal, for another to wear the ring would destroy that magic and render it useless," Leo smiled oddly, "useless, but still very pretty. Here let me put it on you, once you have this you may never fear from being assassinated."

Cesare drew back. Assassination and the possibility of being assassinated was such a common aspect of his life that even a hint of it brought apprehension to him, and it was this paranoia that Leo was praying for.

"Who said anything about being assassinated?"

"Ah, my lord, someone as powerful as yourself must always give concern to the possibility of a dark blade in the night, a poison in a drink from a friend. That is the matter of politics, which a prince such as yourself must be always aware. Here let me put the ring on."

"No! No rush!" The young prince leaned his back against a wall. He had no reason to suspect Leonardo, rather he suspected everyone, even his family – especially his family. "Let me look at it more closely."

He took the ring out of Leo hands and held it up to the light. It glittered in the way only finely made jewellery can, that shine that makes men give up kingdoms and women husbands. The ring was filled with odd shapes and protuberances, arcane symbols rippled across its surface; it worried Cesare that these could hide a poisoned needle.

"Here!" he said after a moments indecision, "you put it on."

"But my lord, the magic will only work for you."

"That's what I'm afraid of. So, you put it on first."

"Alas, my lord," Leo almost grinned as he sought tears, "Should I do this, the magic will be void, all my work will be in vain."

"If you don't do this, I'll have you slain."

Leo smiled sweetly, this was exactly what he was hoping for, and he slipped the ring on his finger, he also managed to release a spring hidden within the sleeve of his cloak that was attached to a chime.

"What was that noise?"

"That was the magic returning to the stars, my prince. Alas, the ring will no longer work."

"Ah, well go make another one," Cesare said, now fascinated.

"Alas, my prince, there are things that even Leonardo can not do twice, the stars have moved in their orbit and it will be a thousand years before they return." He smiled ruefully, "But here, is this not a pretty ring, nonetheless?"

"Well, yes, I suppose," Cesare slipped the ring on his finger and admired its intricacy. "So, hmm." Although in the back of his mind, he suspected he had been duped.

"Now my lord, about the mechanical colossus?"

"I'm not sure if you have held up your part of the agreement."

"We can not argue with the stars, my prince, and you do have your ring of invisibility."

"But it doesn't work."

"It doesn't work anymore, but it would have worked, my prince."

"So you say. Well, before we spend the entire treasury on this mechanical man, lets see you build something smaller, let's say a giant horse."

"Of course, my prince."

"One that breathes fire, soar through the air and pulls a plough."

"Breathing fire that's easy, but pulling a plough and flying at the same time?"

"Alright, just one or the other, but not at the same time."

"Of course."

"And you can do this?"

"I am Leonardo."

"Right," Cesare was so not sure, "and what will you make this beast from?"

"Oh, the usual, pigs bladders, hot air, scraps of timber and some sulphur."

"If that's all you need, then why are you asking for money?"

"I need a vast amount of pigs bladders, hot air and sulphur."

"Alright," said Cesare, "just make me a barking dog."
"How about an engine of steam?"
"What does that do?"
"Makes steam, lots and lots of steam."
"Order me two."

He returned to his studies of the arcane, the profane and possibly the insane.

On the table was body of a model, the dead body of a dead model, a dead model, Leo's model had had an accident. The body of the model was not so much lying on the bench, as encased or wrapped in the wire and struts of another failed experiment in flight.

He had somehow reasoned that since boys have a more animated nature, not unlike the flight of birds, and so it would make sense to use a boy to fly his winged machine. They had launched a peasant lad off the back of the monastery, carefully tying him into a harness made from pigs bladders, bees wax and string. The abbot had questioned the need to launch at all.

"But Leo, Leo," the abbot quizzed, "what if he can't land, what if he can't get back to the ground?"

This possibility had never occurred to Leo, and within half-an-hour he had designed a working parachute and strapped it to the boy's back. The problem with this was that it doubled the weight of the mechanical bird and made it all but impossible for the lad to get any lift from flapping the wings. Unfortunately, Leo had gotten also the decimal point too far to the right on the equations of flight, and the boy had returned to the ground with the grace of a cannon ball. Hence, and as a result, the poor lad had fallen, not flown, straight into the well at the bottom of the building.

"Mama Mia! Mama Mia!" Leo moaned, "this time I really will have to move to Holland!"

There was a groan from the table.

"Zombie!" came the universal cry of the monks as they fled the room.

"It's lives! It's lives!" Leo cried, "I've created life! - Funny I thought I needed a lightning storm."

"Help me," came the groan again. "Help me."

"Of course!" cried out Leo, "he never died, thank God. I thought I was for it."

"Help me."

"Quiet, I need to think."

Leo began to pace back and forth, as he chewed on dried fruit and wondered where the flying machine had gone wrong.

"Help me, please."

"You'll be paid a Florin for your services, now quiet. Leo is thinking," Leo muttered almost absentmindedly.

Before long, the monks and the abbot returned when they realised the boy was not dead but badly needed their aid, they also found they had to impose themselves upon Leo to cut a way into the flying machine.

"Are you mad? It took me three weeks to make that bird."

"Leo, he will die without your help," pleaded the abbot, "he has a soul."

"So did that machine, hmm, mind you if he does die it's going to cost well over ten florins to pay off his parents," Leo muttered.

He swiftly set to cutting out the boy and freeing him from the tangled wreck. With his medical skill and anatomical expertise he was able to set the boy's bones, and more importantly with the head facing the right way. He did however accidentally castrate the boy, and made him at least in the eyes of the church into a woman.

"What's your problem," Leo argued with abbot, " he will sing like a bird, he can't fly like one, but at least he will have as pretty a note."

Soon the boy, with his attached bandages and splints became a permanent fixture in the workshop. Leo discovered the boy made the perfect model, since with all the wires and splints attached to him he could shaped into in position. He also gave the boy a tincture of laudanum to kill the pain and shut him up, this had the side effect making the poor lad addicted to opium.

Two weeks later and just before Christmas, the prince appeared unannounced, as was his nature, at Da Vinci's door. The artisan and the tyrant stood before each other, one a prince of Politics and the other a prince of Art.

"My prince," Leo swept his cap once more to the floor, "how delightful of you to visit as one as an craftsman as humble as myself."

Even the Abbot raised an eyebrow at this one.

"Hmm," Cesare smiled good humouredly, "and what have we here?"

He pointed at the boy swaddled in bandages.

"A model, a poor substitute," Leo said and then joked, "but it's cheaper than a prostitute."

"You're a pretty one," Cesare said, mistaking the lad for a girl, "and why is she all wrapped up in linen, pigs bladders and string? Is she my present?"

Both Leo and the chief abbot looked at the Prince in surprise, but since the last person who contradicted a Borgia was found swinging from a rope at the bottom of a moat, neither of them felt the need to point out the error.

"I was using … her … for a drawing," Leo thought quickly, "you know, one that is anatomically correct."

"Really, hmm, could you do picture of her for me, lets say for a Christmas present?" asked Cesare, though it wasn't really a question.

"The model?"

The boy sat torpid, moaning occasionally, partly from pain but mostly from laudanum.

"Yes, I like her."

"You like the model?"

"I am Cesare."

"Of course, and a painting you shall have."

"Good, and if it is acceptable we will discuss your giant mechanical man."

"Your highness is too kind."

The boy moaned in pain.

"What is she trying to say? Is she an idiot?"

"Why yes, my prince, a village idiot who can only say her own name. Mona."

"What a pretty name for a pretty girl."

The abbot held up his hands to the heavens and was about to speak, but Leo put his finger to his lips to shush him.

So the boy was painted in a faux mountain scene, his eyes wide with laudanum and a strange soporific smile that the prince never understood, even on his wedding night.

In the end, Leo returned to his sketches, his sketches were his pride and joy, and he felt his greatest achievement, because normally when a girl is invited by an artist to come look as his sketches there was a good chance he would get some action – but when Leonardo Da Vinci said come back and look at his sketches, it was a dammed certainty he was going to get some action. As to the giant mechanical man, it never came to fruition there simply weren't enough pig's bladders in the whole of Italy to bring it off.

Einstein and the Zeppelin

In which our intrepid Daedalus flies too close to the Clouds.

Einstein rode his bicycle along the shore of Lake Constance in Württemberg, the crisp morning air opened his mind like a spring day in a tulip garden, and he fell to thinking upon one of the great mysteries of the world. He was wondering about Grammar. It puzzled him no end that words and ideas were made of abstract nonsense, they were not solid or delicious like strudel and yet still they had structure and form, in the same way Space and Time were 'somewhere' or 'somewhen'. You couldn't point at a bit of Time and say, 'then is a bit of time' or 'here is a bit of space', since all you could see here or then was some nothingness. Grammar was very similar; it was not the sort of stuff you could poke with a stick. Yet 'space' like 'words' refused not to exist. Strudel on the other hand was very easy to understand - either it was or it wasn't, and he felt best just after it ceased to be.

As he passed out of some philosophical woods he saw the faultless image of a vast white airship reflected across the very still waters above Lake Constance, hanging as an indifferent cloud to the countryside, it made a mockery of gravity and the rational mind.

"And I thought Grammar was nonsense?"

Count Ferdinand von Zeppelin puffed heavily on a cigar as he stared in admiration at the long missile shape of his second airship - the Luftschiff Zeppelin Two. Investors and mechanics stood respectfully behind him, allowing the great man his moment of glory as the great air ship hung above the mirror waters of Lake Constance. Count Ferdinand felt the moment was perfect, and stroked his moustache thoughtfully, remembering the last twenty years of struggling towards this moment, the endless arguments with officials, investors, war ministry, engineers, the late nights staring at blueprints striving to refine the shape and the engines just a little further than any designer had ever been.

"Gentlemen," he turned to the expectant audience and began a prepared speech - which he had also been working on for years, "today we see a new dawn in the history of the world, for today with our new improved airship we will take the human race higher than it has ever been before, to boldly go where no balloon has gone before - above the rain!"

There was wild clapping and enthusiastic hat throwing.

Amongst the gathered crowd Einstein had wandered in and was standing at the back and also admiring this revolution in flight. He thought for a moment, pursed his lips and in an odd fit of grammatical perversity held up his hand.

"Um, do you not mean, 'to go boldly' and not 'to boldly go'?" he asked innocently. "Is it not an error in grammar, a 'split infinitive'?"

The Count flustered, glared at the interloper, but it was too late for the moment had been spoiled by the question and he knew it. He opened his mouth but no sound came, so far from the instant had he been thrown.

A journalist in the audience spoke up: "It's a common error, even academics misuse it."

"Nonsense," retorted an engineer who had misheard the question, "if you split infinity you still get infinity, and what has that to do with grammar?"

"He didn't say 'infinity'," the journalist threw back, "he said 'infinitive', there are two different things altogether."

"They sound the same, how do you know they're not the same?"

"No,- they don't!"

Einstein had opened a floodgate of trivial conversation with his question and soon the entire crowd was babbling verbs and grammatical terms as if they were still in school.

The count Ferdinand von Zeppelin gaped and tried to interject, futilely raising his hand, but the flow of nonsense continued on unabated for several minutes. The audience he had gathered to witness what was a truly historic moment, were now more interested in the difference between ordinal and cardinal numbers, or whether a past perfect predicate was imperfect or impacted.

"Silence! Enough of the grammar already!" he finally screamed, "Zeppelin! Me! Airship!!"

Then turned around and strode off to his airship chewing on his cigar, so annoyed he could barely contain himself. It never occurred to him that split infinitives did not exist in German.

They all boarded the ship and it clattered into a blue sky above the lake, an improbable bird soaring an impossible flight. The earth fell away leaving the passengers breathless at the ever-widening spectacle of the countryside of Württemberg. Red tiled houses, flat fields and lowing cattle scattered across the fields and about the lake, while in the distance the chiming of church bells was echoing from steeples and towers. Whether the migrating flocks of storks flying gave any thought to the giant airship was impossible to decipher in their honking, most likely they looked upon it as no more than feature of the air-scape like a cloud or the sun – it was certainly nothing compared to the thoughts and emotions of the crew or passengers of the rising airship.

"Is that a cow?" one journalist asked another.

"That's a horse," said the other.

"Looks like a cow."

"It's to do with perspective."

"I'm sure it's a cow."

"Why is it on top of a house?"

"That's a barn."

"Why is a horse on top of a barn?"

"It's not, that just it's perspective."

"You're telling me, that looking at a cow from the perspective of an airship makes it look like a cow on top of a barn?"

"No, that is a horse."

"Gentlemen!" von Zeppelin cried in exasperation, their babble was nothing like he imagined their response would be. "We are not here to discuss cows or barns! Or farm animals of any kind! We are here to rise to the top of the world, beyond the clouds! To almost reach out and touch the stars!"

"Will we get wet?"

"What?"

"Clouds are wet."

"No, they're not!"

"Course they are, that's what they are made from ... wetness."

"Wetness is what they become," retorted an engineer, "they are made from gas and smoke. It's only after they fall from the sky they become wet. Dryness is what holds them up."

The count chewed his moustache angrily, and began to wonder what might be the legal implications were from dropping off people as ballast in midair.

"Gentlemen, please, this is an historic moment," he protested once again.

"I didn't bring a mackintosh."

"Will umbrellas work at this altitude?"

"No, rain falls, but clouds rise. You would need an upside down umbrella."

"Couldn't I hold it upside down, anyway?"

"How? I mean you'd be standing on it."

"Under the airship."

"Ah. Clever."

Again this was too much for the Count and again he strode off to the other end of ship to puff angrily on the remains of his cigar. As he stared at the rolling hills and endless farms across Germany, he started to think about migrating, anywhere, anywhen, somehow, someway and watched wistfully the flocks of storks. The dirigible continued its ascent, driven by the gigantic gas cells of hydrogen, which enveloped a rigid aluminium alloy skeleton. He rested an elbow and wished he had never brought the spectators aboard. It was at that moment that Einstein appeared at his side and also stared at the wonderful countryside. The airship's bridge was a glittering framework polished aluminium struts and glass dials.

"Ah, it is like being an angel rising up to heaven - only without the feathers."

The count turned and smiled, here was a kindred spirit, perhaps an annoying kindred spirit, but at least he occasionally said the right thing.

"Yes, that is what I feel too."

They watched the horizon creep more distant, and soon the first wisps of cloud began to form about them.

"Rain?" They heard from behind them

"Ach, not again."

The count glowered and considered returning once more to the landing field.

"What must it be like for God to stare down at us all, through the clouds, to sprinkle rain and whisper of eternity to us?" He muttered.

He turned and faced the crowd in the cabin once more,

determined to enthuse them with the glory of the moment with another of his prepared speeches.

"The day will come!" he yelled, "when we will have a steerable train of airships, an Orient Express of the sky, which will roam the whole wide world. A luxury airship, a fabulous skytrain."

"Will it have a cloud-catcher?" asked a journalist.

"What?"

"Like a cow catcher - only for clouds."

"No. It will not have a cloud-catcher," von Zeppelin fumed, "but perhaps it is time you helped me test the parachute."

"What happens if we run into a cloud?"

"Nothing will happen! Clouds are made from gas!"

They entered a cloud at that moment.

"See! Nothing happens."

"It really is wet, who would have guessed that one."

"I did," said an engineer.

"Yes, but besides you, who would have guessed it."

"Anyone who thought about it. It's pretty obvious."

"No it's not, it's not obvious at all. Rain is wet, it falls, so how can this wet stay up?"

"Maybe it's not rain, it's dry water?"

"You mean snow?"

"Arrrgh!" Cried the count and stormed off for the third time to hide in rear of the ship's bridge. As they left the enveloping cloud brilliant sunshine shot into the cabin blinding them, and the steady chopping of the engines droned on, but the guests had followed the count and stared out his window.

"Clouds, we are in the land of the clouds!" he sighed, once more trying to enthuse them with a sense of historical importance.

"Is that a cuckoo?" asked a reporter.

"What's cuckoo doing at this altitude?" came an engineer, "I thought they were arboreal?"

"It's a stork! It's a migrating white stork!" the count finally broke, "you stupidly insane troop of faeces throwing monkeys! It's a damned stork!"

"It's headed right for us."

"So it is."

"Do they bite?"

The bird on its long flight from Africa had fallen asleep on the wing, and was unaware of the approaching collision. Einstein who was still at the front of the bridge and still admiring the ship's glittering dial and controls, was also unaware of the dangers of midair crashes, and had no idea he was about to rear-ended by a slumbering stork. The count and his audience stared in amazement as the heavy bird dashed inside the nacelle, shot past them and flew straight into the back of Einstein's head. Who fell forward and grabbed a cable as he fell outside the gondolier and dangled at the end of a rapidly lengthening control wire.

"Gott im Himmel!" He screamed appropriately as he swung through the air, "I'm Dezeppinstrated!"

Unfortunately - and most importantly - the cable was connected to both the ship's ailerons and motors, which dragged from their normal positions, forced the airship to sharply pitch up. This in turn forced the crew and guests to slide to the back of the bridge and fall in a writhing screaming mass. Einstein swung in a wild arc under the airship until he came full circle and fell through the rear portal of the gondolier, to the astonishment and terror of the swarming mass of arms, legs and assembled passengers.

It several minutes for one of the crew to crawl back to the controls and bring the aerial leviathan back to a horizontal plane, but in that time the damage had been done and they discovered the airship had lost the ability to descend.

"Gott im Himmel!" screamed a journalist, who should have known better, "we are going to crash into the Moon!"

"Don't be ridiculous. The Moon is a balloon, we will bounce off it."

"The Moon is not a balloon, this is a balloon."

"No, this is a airship," another engineer interrupted and pointed at von Zeppelin. "There, ask the count, he should know, it's named after him."

"It is?" asked the journalist. "You know I never made the connection. Pity, we are all going to die after we run into the Moon. I'll never get to use it as a headline."

"You know," said Einstein thoughtfully, "you could write a message and attach it to the parachute, then drop it to earth. That way..."

He never finished as there was a mad scramble for the ships only parachute, in the wild melee it was torn to pieces and the headline was never written.

As the airship continued its ascent, the air became cold and rarefied.

"How long till we hit the Moon?" Asked one of the passengers.

He was soon hit.

"Ow!"

"Somebody has to climb out to the motor and steer by hand," said the count finally dealing with practicalities. "I would volunteer, but I fear I am too fat."

In turn, they all gave medical reasons varying from fear of ducks to an undying dread of getting a rope burn, ephebophobia or the irrational fear of adolescents gaining social emancipation and a fear of grandmothers with stainless steel knitting needles. Eventually they all looked at Einstein doodling an equation on the airframe, he looked up, smiled wanly and knew it was one of those days.

"I have Zeppelinastraphobia," he pleaded, but an irrational fear of being hit by lightning at two thousand feet while in a runaway airship simply wasn't convincing enough.

They tied a rope to him. First they tried to tie it to his neck but he pointed out this may actually kill him. They pointed out that was not their concern, until he mentioned he couldn't adjust the engine if he actually was dead. Next they tied his feet together and he fell over, quickly followed with tying his hands. None of which had the desired effect of securing him. Eventually they listened to his idea of using a harness that was hanging on the wall next to them.

He swung hand over foot across the airframe, giddy with fear. With each swing the airship he crept further from the relative safety of the gondolier and closer to the churning motors.

"Go around them!" yelled out the Count, "you must get to the controls at the front!"

It was at this point he realized it was amazingly dangerous. The spinning propellers impeded his path and there was no way round. He clambered back to the gondolier and explained it was impossible to do without losing large parts of his anatomy and that he was very attached to large parts of his anatomy and wished to remain attached.

"Ach of course!" Von Zeppelin commiserated and they dragged Einstein to the front of the bridge. "If we throw him - fear not" He laughed as Einstein drew back in alarm, "You will still have the rope attached. You can swing underneath, like mister Tarzan as you did before, only this time with a longer rope past the propellers and grab the rear of the airship. Manipulate the controls from there. It is our only hope."

"But what if the propellers cuts the rope?" Einstein's eyes were wide with fear.

"Ach good point, I know we will use a wire instead of the rope, the propeller can't cut that. Not to worry," patting him on the shoulder, "you can have the parachute as well."

"But they just tore it to pieces."

"It's only for psychological reassurance - I wouldn't use it."

Thereupon they switched the rope for a steel wire and tied the parachute to his chest. With a wild huzzah they threw him from the front window of the airship. He flew in a graceful arc screaming.

"Twice!"

All did not go to plan. For not only was the wire so long he overshot the end of the airship but they passengers found it impossible to grab the steel cable and pull it back. Einstein fell back and began to pendulum back and forth beneath the airship. Thereupon the matter became even worse as the ragged fabric of the parachute blossomed open to catch in the wind and held him as an airbrake beneath.

"Pull me back!" Einstein screamed in terror.

"We can't! We can not grip the cable!" came von Zeppelin's answer, "besides you have the correct aerodynamic form! Stay there! This may work!"

"I want my bicycle!"

Einstein's weight and the pull of the ragged parachute had succeeded where their original plan had failed and was redirecting the nose of the airship downwards. It was a minor correction but when you are trapped in a runaway balloon at three thousand meters with a broken control system and a group of screaming passengers. Your only hope is with a passing angel. After a while the passengers settled down for coffee and biscuits and their conversation of grammatical nonsense, while beneath them came the mournful cries of Einstein as he swirled through clouds and ran into migrating storks.

At the front of the gondolier the Count assessed the situation and discovered they had flown halfway to Switzerland and were approaching the ominous Allgäu Mountains. In their path rose a glittering line of snow covered shark teeth, jagged and triangular peaks gnashed against clouds and blue sky. Fretfully he chewed on his moustache, this simple flight was turning more and more into an alpine expedition with every minute. He looked down at Einstein and marvelled at the capacity of his lungs to scream in terror.

"You are doing very well," the count reassured him, "do you want a coffee?"

Einstein stopped screaming, "Italian or Bavarian?" he asked.

"Italian."

"Yes, pull me up. I take sugar."

"No up. That parachute is our only hope. I'll send it down."

The count tied a slipknot to a thermos and let it slide down the cable. Naturally it shot down the wire with increasing velocity and painfully slammed into Einstein's fingers who began screaming for a moment then stopped, as he called out:

"No sugar?"

"All out."

The screaming began again.

A storm began to rise about them as the clouds they were travelling with began to crowd against the base of the Allgäu mountains and were pushed higher and colder, as slowly the dark swirling of the cloud changed to thunder and the ill-omened crack of lightning. Gradually the air became charged with electricity as everyone felt the hairs on their head begin to rise and stand upright, and through the murk they saw it was impossible to distinguish the mountain from the boiling storm clouds. The tension exploded with a sudden, mind-numbing explosion as all around them chain lightning raced across the billowing clouds. Ball lightning bounced about the skin of the airship threatening to cut holes through its aluminium surface. Suddenly there was a sharp down current of air and the airship plummeted through the swirling mists and downpour at stomach emptying velocity.

"Hold on!" cried out the count.

"You think we are letting go?" screamed back the crew in various forms.

"You down there!" the Count cried down to the hapless physicist. "When we get close to the ground, secure the wire to a tree or something!"

"How do I do that?!"

"Improvise!"

This was met with a loud continuous wail.

"Count Zeppelin!" one of the crew yelled in alarm and pointed ahead of them, where through streaming rain they saw the pinnacle of a church steeple cut through the fog like an iceberg slicing open a cruise liner. "Church dead ahead!"

"You down there! Grab hold of the bell tower!"

"Our Father who art in Heaven – oh wait, I'm Jewish!"

Einstein did, but released himself from the harness at the same time.

"No!" The count cried futilely, but it was too late and the great airship sailed on into the storm.

Einstein lashed himself to the bell tower and was discovered the next morning fighting off the white storks that had nested there.

Newton Illuminated

Science begins with Misfits and Upstarts.

Along the Backs of the college, Newton could make out the figure of a bargeman leading his draught horse through the river Cam and pulling a barge behind it. Years before, Newton had insisted the college refuse to allow bargemen access to their bank, and the longboat men retaliated by building a submerged path in the canal itself, to allow the horses to pull their great loads. To this day, the bargemen still stuck their fingers up in the air if they saw Newton at the Backs of the colleges.

"Up yours mate!" came the customary greeting of one of the river brethren when he saw Newton standing at his window. Newton frowned and walked to his other window, which faced the Great Court.

At this point in history the Great Court was little more than a muddy field, and in it he saw the inevitable cows with flowers woven between their horns as they lowed in the morning sun. Sheep ate grass and baaed like students as they gambled about in the morning mist at the centre of Trinity College.

Newton picked through his mail by the door. It consisted amongst other matters, of bills, letters from cousins Norfolk instructing him to sell off his apple orchard, an offer of marriage, and an account from the colonies of New England of the Plymouth Council to regulate the length of whales. As he came to the last letter he gave a sudden start and called out for his manservant Hedby.

"Hedby!" Newton looked at his door, "You wretch! Where are you?"

"Over here, sir," came a voice from behind him, "behind you on the couch, I'm mending your socks."

"Ah, there you are. How long have you been there?"

"About a week, milord."

"Hedby, I have a letter from his Bishop of Nottingham informing me of his visit."

"Yes milord, I know."

"How can you possibly know?"

"Well sir, I have a look at your letters, so to speak, before I bring them in. You never know when some snippet of gossip might come in useful."

Newton lowered his head and gave a menacing stare at his servant.

"You read my mail?" He said somewhat angrily, "how long have you been doing that?"

"Since you told me to make sure you got no useless mail, milord."

Newton flustered and twisted about in his chair, puffs of talcum powder cascading from his wig.

"That doesn't mean your supposed to read them, just the bit on the back where it says it's from to make sure it's not needless mail. Have you no idea of privacy?"

"Ah, good thing I can't read too well, ain't it milord?"

"What? But if you can't read them, how do you know what's in them?"

"I show them to Mister Bainbridge, the bell ringer, he was a teacher before he came down in the world."

"Damn your eyes Hedby!" Newton shouted and stood up, "you're not supposed to read my mail! Zeus in Heaven! And you're certainly not supposed to show it to anyone else!"

At that point there was a knock on the door.

"Isaac?" came a steady voice, "you there?"

"What is it now?!" Newton thundered.

"Its me, Charles the Chancellor of our University, may I come in?"

The head of Charles Seymour appeared at the door, it was such a handsome head one might almost imagine an Italian artist carving it out of marble and then God endowing it with life just to annoy the artist. Like Newton, he was a Trinity man, which is odd as both of them were Unitarians. Newton fell over himself letting his paymaster into the room.

"Charles! Do come in," Newton swept his wig to the ground, absent mindedly thinking it a hat.

Charles Seymour, who had never seen Newton make a joke in the entire time he had known him, believed Newton had finally developed a sense of humor. Laughed out loud at the absurdity of it and in return he swept his own wig to the ground. Whereas Newton, not realizing he had used his wig instead of a hat, thought Charles Seymour had gone stark mad.

"Yes, yes," he said, "do sit down."

"And take a load off my wig?" quipped Charles as he twirled his wig about his fingers.

"If you wish," Newton regarded with Charles with some concern.

The smile froze on Charles' face and with a sigh he returned his wig to his head. "Well, old chap, I've come about a matter that is of no little importance."

"A pay rise?"

"Nothing so prosaic, no old man, this is very serious matter," frowned Charles, "and given your background in crime-sleuthing I thought it best to approach you in the matter."

"Aha!" Newton exclaimed, forgiving him for his sudden attack of lunacy, "just when I felt I was getting rusty in the weather of natural philosophy! Well old man, how can I be of assistance?"

"As you know, I am amongst other duties besides organizing the colleges for the inter-college games."

"We have games?" Newton asked in surprise.

"Yes, the yearly rowing competition amongst other events. Surely you must have noticed it from your rear window. It's barely thirty yards away."

"That? I thought those bargemen."

"No, those bargemen are in fact bargemen." Charles frowned. "In fact, we only hold the rowing competition as way of getting back at them for using the river without our permission."

"I see."

"The thing is old man, there has been a theft from the Great Hall."

"At last! Something to sharpen my wit on!"

"Newton, I'm glad you think so," Charles mused. "Although, I can't help but wonder why you don't engage yourself upon science, as you used to do? You are after all, the world leader in the matter."

"Charles, the world of natural philosophy is like a giant book that anyone can read. I have read as much of it as God will allow in our age. To peruse the tome of God any further would be an error of hubris, which even I would not wish to risk. Now then, you have a crime for me? Delightful. Hedby, go make a pot of tea for our guest."

"With or without milk?" Hedby asked mournfully.

"With."

"Oh dear," Hedby's face fell and he picked up a bucket from the table and walked out to the Great Court.

"Now then Charles, give me the facts," Newton sat himself upon the windowsill, "merely the facts without embellishment."

"It's like this," said Charles as he stood besides Newton to watch Hedby chase a cow. "As you know, each year we have a celebratory dinner at the Great Hall on occasion of the inter-college games…"

"No, this I did not know," interrupted Newton, "why have I never been informed of this before?"

"Oh yes, that's right," Charles looked embarrassed, "you may not have got an invitation."

"Why on Earth not?"

"Well, you see old man," Charles looked out at the court, hoping for inspiration, "the whole affair is all rather a bit silly and … well. No one thought you went in for that sort of thing, being all serious that you are."

Newton who had a reputation of never laughing in his entire life and of being permanently angry, gave Charles such a look of wounded pride it would have done justice to a lion after a troop of baboons had flung spoiled fruit at him.

"Are you suggesting I'm not funny?" he shrilled.

"Oh no, well … err … yes." Charles began to sweat and edged away from the window. "Well, you are you, you know, and well you didn't get to where you are, by playing boyish pranks. Did you?"

Newton wrinkled his lips and half-nodded his head in agreement, then turned back to watch his manservant.

"Hmm, quite, well go on." Newton frowned then yelled out, "Faster Hedby! We don't want the milk tomorrow!"

The cows galloped about the court as Hedby, bucket in hand, trailed behind them, as students clapped and hooted as they lay about on the grass.

Charles breathed easily and walked back to fireplace to sit down.

"Well, the facts are these the Great Hall has been entered into and a table has been moved," Charles said rapidly.

"Yes, go on."

"That's all, old man."

"No really, what was stolen?" Newton looked annoyed. "Someone murdered, a mysterious letter. Those are facts."

"Well, actually, the only crime is that a table has been moved."

"Charles!" Newton shrilled. "That's not a crime! That's a cleaner!"

"I assure you Newton, all the porters and cleaners have been questioned and none of them appear to be involved. No, I fear it was due to an outside agency."

"A table?"

"Well it does weigh eight tons."

"Oh, that table," Newton leant back his head stared at the ceiling. "The High Table of Trinity."

"Quite."

"Well, it's nothing more than a student prank, Charles. Doesn't take a professor to recognize a student's hand at this one."

One of the cows had had enough, gave a tremendous bellow and began to chase Hedby instead.

"Faster Hedby!"

"I'm afraid it's more serious than that, old man."

"Ah, there is something you're holding back," Newton turned and faced him, sitting himself on the windowsill.

"The key for the Great Hall which normally resides in my chambers was still in the room, and most importantly the door was locked."

"So presumably you were forced to use a locksmith to open the door."

"Quite."

"And at first you thought the key had been stolen, but upon discovering it within a locked room, and I seem to recollect the Great Hall is almost a fortress, you decided that it was not a theft but merely an entering."

"Quite again."

"I think...," Newton leaned back on the window sill and fell out onto the Great Court, "that … !"

Charles rushed to the window as Newton popped back up.

"Quite all right, Charles, merely an excess of thinking," Newton brushed himself down in annoyance. "Well, let's go see this room and its great mystery, shall we?"

"Shall I?" Charles looked down at the low windowsill, which reached almost to the ground.

"Yes, it will be quicker and we can go straight across the quadrangle." Newton offered his hand to help Charles through the window, when behind him came the thundering feet of Hedby and the cow.

"Milord!"

Hedby and the cow were charging straight towards him.

"Damn your..." Newton yelled, but never finished as he dove back in the window.

Charles and Newton collapsed on the floor.

"Well, perhaps," Newton said uncomfortably and they helped each other up, "we might take the corridors instead."

The Great Hall or King's Hall was a little known part of Cambridge that the scholars retreated to when given the chance and was normally kept locked. Hidden at the Backs between Trinity and Saint John's colleges it had originally been part of a smaller Michaelhouse hall, which under the orders of Henry the Eighth, had been combined with several other buildings to create Trinity. The room itself was large and luminous with splendid carved panelling and arching delicate internal buttresses. In the centre of the room an enormous table dominated the space, it was around this table that the high dinner of Trinity were held each month with dishes such as roast swan, fried hedgehog and pâté en terrine of black eels which were a particular favourite of the scholars. The surface of the table was ornately carved with medieval heraldry and mythical bestiary, rimmed with a Latin inscription that ran the entire circumference of the table.

"See, Newton," Seymour pointed at the giant table, "It has been rotated a full one hundred and eighty degrees."

"Rotated? You bring me here because of a rotated table?" burst out Newton.

"A rotated table, in a locked room with the key for the door dangling on the high chair."

"No offence, Charles, but it's hardly a crime."

"On the contrary old man," Charles picked up the keys. "I hold it to be a crime against tradition."

"As you wish," Newton frowned as he paced about the table, "well let's see what the room has to say."

Bending down he ran his hand along the edge of the table looking at the fine layer of dust that covered the floor. He looked up at the ceiling and then jumped up on side of the table to look at the windows.

"The windows are sealed," he commented after several minutes of bending, jumping and clambering about the woodwork. "Lead sealed, stained and glazed, impossible to open without shattering the glass. Most odd."

Seymour helped him down and they went back to the table, Newton looked at the ceiling.

"You will note the carving of a rose flower on the ceiling in the center of the room. In Latin this "Sub Rosa" is a coded sign, meaning this is a place of a secret society. Although of which society I have no idea," he said pointing at the ceiling as if he had suddenly discovered a gargoyle reading a book.

"I thought that was there because of the War of the Roses."

"That too," Newton frowned at not spotting the obvious, "and you say the key was definitely locked within the room."

"Quite."

"Another key?"

"What would be the point in taking my key?"

"Ah," Newton mused, "I see now why you came to me. It boggles the mind."

"What does a boggled mind look like?" quipped Charles.

Newton regarded him dourly.

"There can really only be one solution."

"I'm all ears, old man," Charles flapped his wig cheekily.

"They are still here."

"Who are?"

"The people who moved the table."

"Newton, wouldn't we see them?" Charles looked about the room with a perplexed look, and then ducked his head down to look under the table. "I mean there's just you and me, are you proposing they are invisible."

"Yet it is the logical conclusion," Newton said simply and shrugged.

"Right," Charles said ironically, "maybe I should ask the gardener's for their opinion instead."

"Damn your eyes Charles! There can be no other explanation!" Newton thumped the table in annoyance, and then to his surprise the enormous slab of oak rotated half an inch.

"It moved!" Charles almost shouted.

"No question of it," said Newton as he banged the table again and once more it moved fractionally about its axis.

"How odd I've never noticed this before."

Newton dropped to his knees and peered under the table,

examining its legs.

"Look there in the centre," he said, "I'd swear that's an axle, and here the legs have stops on them to prevent the table from moving but at present it appears they have been released."

"But Newton this is simply amazing."

"You think this is amazing, you should try some of Hedby's cooking."

"But what can it mean?"

Newton stood up and began to pace back and forth. After an hour of this Seymour went off for lunch leaving Newton to his pacing. He came back in an hour and to his surprise the mathematician was still pacing back and forth.

"Isaac, have you any idea what it all means?"

"What? What?"

"The spinning table."

"It is a giant wheel for a giant pulley."

"Never."

"I assure you it is."

"But what on Earth for?"

"To open a large stone door which is hidden in this room."

"Why can't we see it?"

"I suspect the mechanism is old, and perhaps a rope or gear has broken," Newton held his finger up and he continued pacing, "whomever used the mechanism is probably trapped beneath us. There really can be no other explanation."

"Brilliant."

"Yes, I know," Newton preened. "Now examine the Latin inscription about the table."

"Valde rota of polus verto quod nos verto subter supter is," Seymour read aloud.

"The vast wheel of the heavens turn and we turn beneath it."

"But won't that mean, they are still trapped down there? And will they not be running out of air? Great heaven's what are they doing down there anyway?"

"Seymour there is not a moment to be lost! Send for the Navy!"

"Don't you mean navvies? Workmen?"

"Ah yes, them too."

A horde of porters, gardeners and even students were brought into the great hall, given shovels and picks to did up the flagging in the Great Hall. The digging went on for several hours without a secret passage or gearing to be seen. It was only when a retired professor of theology happened to be passing the door and asked them why they were destroying the swivel table for the dinner set, did they realized it was a wild and pointless dig. They all stopped and looked at Newton. It was as if a goose had gotten up from the table and asked for its head back in the middle of the third course.

"Newton, I'll have a word with you," Charles crooked his finger.

"Damn my eyes, Charles, what other conclusion could I draw."

"One that involved a lot less digging," Charles frowned at Newton disappointingly, "Nevertheless, I'm now convinced that there is something more to this than immediately presents itself to the mind. I want you to continue this investigation after you have replaced all the flagging."

"What?!"

"Good day to you," he said and left, closely followed by the entire hall of staff and students who walked out noisily dropping their shovels at Newton's feet.

"It's not my fault!" he yelled helplessly.

Three days later, he and Hedby finally put the last of the flagging stones back in its place.

"Hedby I've been thinking."

"Yes milord, I expect you have, being Sir Isaac Newton and all."

"Very droll," Newton frowned, "I want you to listen to what I have to say and give me your opinion on the matter."

"So glad of you to take me into your confidence, milord."

"That's because nobody else is talking to me."

"Yes, milord."

"There is a means by which the key arrived on that chair. If someone had taken it from the Charles' office, ascended the roof of this hall, lifted a tile from the ceiling and finally used a string and a hook to lower the key down onto the chair."

"Or a monkey."

"Damn your eyes Hedby," Newton exploded. "There are no monkeys in England, they only exist in tropical Africa and the colonies of New Spain."

"One of the students has one."

"Why in blazes didn't you tell me before?"

"Well, milord never asked his humble servant."

"So the monkey escaped from the student room and goes for a wander about the roofs of the college, finds the keys, carries them about until it drops in here. Zounds I wish I had known this four days ago."

"That is brilliant, milord if I might be permitted to point out a slight problem with your reasoning."

"If you must."

"Well, milord, given that the monkey is dead, stuffed and sits in a glass cabinet - how did it go for a wander?"

Newton fumed for several minutes.

"Hedby, from now on try not to think."

"Yes, milord."

"And Hedby I'd like a cup of tea, with milk."

"Ooh, not again."

Two days later a strange tramp appeared at the Backs of the colleges. With a long straw dangling from his bottom lip and a mass of vivid red hair bursting out from underneath a large straw hat, which he had clamped upon on his head to keep the wig in place. He carried a large ash shepherd's crook and walked with a stoop.

Newton had disguised himself as a sheep farmer.

He had done to this to discretely ask questions amongst the commoners which would have been impossible otherwise. On the bank of the river he found himself talking with one of the river bargemen, but what he had neglected to hide was his voice.

"You there," he called out to the bargeman and he coursed up the river leading his draught horse. "You there, fellow, I would have words with you, if I may."

"Wot the bleedin' eck you on about mate? You some bleedin' toff with a sheep stuck up ya arse?" The canal man laughed and continued down the river.

"No, dear fellow I would have words with you."

The bargeman stopped and stared in amazement at this bizarrely eloquent shepherd.

"Ere," the boatman said disbelievingly, "where are ye sheep?"

"Ah, now that is a good question. However, I would like to ask you some questions, if I may?"

"Ere now, ow come yee talk so lordy for a shepherd?"

"My good fellow, would you please let me do the questioning, or we will be here all day."

"Well, if you insist mate."

"I am conducting inquires to a mysterious break in up at the college."

"You," the bargeman stared disbelievingly, "a shepherd? Right."

"I want you to come with me to the hall, perhaps you may assist me in this inquiry."

"A shepherd?"

"Damn your eyes, stop harking on about my profession. Now will you assist me or shall I be forced to call the porters?"

"A shepherd is going to call the college porters? Right. Listen mate if I get any more lip out of you I'll bleedin'…"

"I'll pay you half a crown," Newton gave up trying to browbeat the bargeman.

"Sold."

They entered the hall; the bargeman was more interested in this strange chimera of a shepherd than he was in the university until he saw the splendour of the Great Hall.

"Oh' my gawd!" he exclaimed in wonder of the stained windows and the carved furnishing. "Oh' my bleedin' gawd!"

"Are you in pain?" Newton had no idea what the bargeman could be experiencing.

"Oh' my bleedin' bleedin' gawd!"

"What on earth are you raving about man?"

"Is this like where they keep the Crown Jewels?" The bargeman even though he had travelled the length and breadth of England had never seen a grander room than lavatory out the back of the Duck and Badger Inn.

"Stop this blathering," Newton rose to his full height and towered over the bargeman, "there has been a serious crime perpetrated here, and I have asked you here to inquire if you or any of you or your barge brethren might know anything about such an occurrence?"

"Wot' a bleedin' shepherd is investigatin' a crime?" the bargeman pointed at Newton's shepherd's smock and crook.

"Damn it, man! Someone has stolen the keys to the Great Hall, entered and locked the doors, then mysteriously vanished. This is of considerable importance!"

"Wot it closes like this?" the bargeman pulled the door shut, its spring lock snapped shut and accidentally locked in Newton. From outside the door, Newton could hear the bargeman laughing. "You bleedin' mean the wind blew the door shut?"

"Damn your eyes!"

Einstein and the Dyke

In which our hero sinks Holland and discovers the North Sea.

The canal was filled with the sounds of drunken captains and their crew singing in the Dutch New Year. A chill wind swept along the frozen river, picking up flurries of snow, twirling it about like dancing Nordic sprites. Tall warehouses loomed over the canal, overflowing with the promise of eastern spices, northern furs, English wool and even more inebriated Dutchmen. Few people could be seen only heard, the cold was so intense that the only reasonable thing to do, and the Dutch being very reasonable people, was to shelter within whatever building, barge or home available. The ice and snow covered everything, the canals, the barges, the warehouses and even a certain patent clerk from Switzerland, who was there visiting his good friend Hendrik Lorentz - one of the very few people, in those days, who recognized Einstein as scientific colossus.

Einstein looked down at the map, he wasn't sure if it was the streets fault or his own. What was even worse, the map made as much sense upside down as it did right side up. Hence, he was becoming a fixture in the snow, as he tried to decipher symbols so arcane it made the Mayan language look like French in comparison.

Lorentz had invited him to spend the New Year with his family, and the map was of Lorentz's design, unfortunately Lorentz had also sent Einstein a diagram for an electrical machine that would have not been out of place in a movie with Boris Karloff.

Einstein was also puzzled by the houses nestled on the road itself - as he thought them to be. For in truth, the barges in the canals were so covered and bound by ice that anyone unfamiliar with their true nature would have been just as confused. After some further minutes of consulting the map, he decided he must ask for directions from one of these houses where the singing was heard, and walked across the frozen street to knock on one of these peculiar houses. Since, he also did not understand the true nature of a frozen canal, the result was as predictable as the end for an escaped penguin from London Zoo attempting to flee via the M1 motorway.

Before Albert had gone half way, the ice collapsed beneath him in a debacle of ice and water that fills every Arctic explorer with dread.

"Gott in Himmel!" he screamed as he vanished beneath the icy waters.

Fortunately for him, and the reader's necessity for entertainment, one of the sailors aboard a barge heard the yell, and knew from long experience that someone had fallen in. Einstein was quickly fished from the canal, amidst much laughter and banter from a drunken crew.

The Dutch were, as always, one of the most benign and benevolent people the world has produced, never a harsh word, even their spoken language seems to gently caress the ears... no wait, I'm thinking of the Swedes - the Dutch character is in fact exactly like their North Sea environment, as harsh and quarrelsome as a raging Force Nine Gale in the Zuiderzee. For a thousand years of building dykes, only to see them destroyed again and again by the North Sea, had created a society of very angry Dutchmen, and it was into this milieu of friendly drunken yet latently aggressive seamen that Einstein was thrust.

"So! My friend!" The captain thumped Einstein good-naturedly on the back, helping to warm as he stood shivering before a coal fire in the crew's quarters of a barge. "A long way from home, you are. Drink some schnapps."

Soon Einstein was comfortably ensconced with them, singing songs of which he had no idea, and swapping jokes of which the crew had even less idea.

In the morning, Einstein awoke to the sounds of water thumping against the side of the barge; he lay there in a cramped bunk moaning and holding his head, wishing all New Years Eves were against the law. It wasn't till after eleven that he was able to stumble his way to above deck and saw in front of him, and to his surprise, that vast and mercurial body of water called the Zuiderzee, and to his left endless windmills, villages and dykes sailing back towards Amsterdam.

"Hello my friend. It is a good morning we are having." The captain smiled, and handed him a scorching mug of very good Dutch coffee, then looked quizzically at his guest. "I thought we had left you back in Amsterdam. Ach, no matter, we can put you off in two days, we sail up the coast, you can catch another boat back to port. More Schnapps?"

Einstein threw up.

Soon, however, he found himself enjoying the breeze, the sights of Holland were truly delightful, and he did not mind the day's excursion. The enormous spinning clocks of the windmills, lazily rotating in the bright winter sun were a delightful distraction. The sailing barge itself turned out to be a marvel, as it surged through the waters, its great sails bulging out in the winds, like a heralds banner announcing the coming of the king. He watched with no little amusement, the young men of the canals impressing the girls with the jumping the canals using a barge pole and occasionally sliding down through the ice. Yet, it was the skating that thrilled him the most, the children shooting along the canals, oblivious to the icy waters beneath and indifferent to the possibility of crashing through.

It wasn't until the second day as they approached the headland along the coast, where he was to be dropped off, that the peculiar nature of Dutch belligerence surfaced as a minor revolt occurred amongst the crew. This, while it had nothing to do with Einstein, nevertheless impacted greatly upon his proposed departure.

A brawl erupted between the captain and the first mate over the identity of the ship's mascot. With the captain insisting it wasn't really a cat and the mate adamantly maintaining that it was. Soon, and to Einstein's astonishment, the entire crew joined in using fists, belay pins, buckets and the ship's cat high tailed it up the rigging. There was a short intermission as they stood about bleeding, drinking gin, looking for the cat, before starting up again – all the while Einstein stood back, jaw agape and wondering if it was safe to join the cat.

The fight continued off and on for about an hour, until with a tremendous crack they were thrown forward to the deck. The barque first came to a shuddering stop then tilted dangerously up into the air, with sliding everyone back to the poop deck.

Einstein's "Gott in Himmel!" was accompanied by many "Godverdomme!" and "Zoon van een hoerige kameel!" as the entire crew, cat and Einstein collapsed in as a ballast of water, people and sails, and the barge began to sink.

"Abandon ship!!" came the cry of all and sundry, as the crew uncovered the ships only lifeboat and prepared to lower it over the side, until Einstein pointed out to them the bow had in fact come to rest on the shoreline.

The barque was resting on the lee side of a dyke, not too far from the sea, and was busy settling into the mud. The crew clambered forward to survey the damage, whereupon another brawl broke out as to who was responsible, who would salvage the cargo, and somebody had to get the ship's cat out of the sails before it blew away. This third and final mêlée continued until the ship sank, the crew dragged themselves on land, and 'Saki' the ships moggy had swum away.

"See! I told you it was a mongoose!" The captain cried as he pointed out the sinuous motion of the cat as it disappeared beneath the waves. "I bought it in Borneo! I should know!"

As a light snow began covering the combatants with white cloaks, the ensemble trudged off along the dyke, the captain occasionally pummeling the sailors, and the crew pointing out that not only had the ships papers and their contracts had gone down with the ship, but there no longer was a ship and he was no longer a captain. This only infuriated the captain more, and he started beating upon them with greater vigor until they picked him up and threw him in the icy waters. Whereupon he climb out greatly subdued and they continued through the worsening weather.

They came to the ruin of an abandoned windmill, it's sails spinning ineffectually in the rising wind, and elected to stay the night there. Dusk was soon falling and they had no hope in the dark of making it back to any of the villages they had passed coming downstream. They huddled about a cast iron stove, and burned the German's clothes to keep warm.

When Einstein had been reduced to his underwear, he began to worry.

"There is nothing for it, we must wait here for morning," said the captain.

There was a vigorous nodding of heads amongst the crew and assorted "Godverdommes!" as they all looked at Einstein shivering in his flannels, and since this was the only thing he had left to burn, he backed up against a wall and defiantly raised his fists.

"We must also get some wood, or we will die of exposure like our guest," said the captain, which brought chortles from his crew.

They set about tearing down parts of the windmills walls,

slowing opening portals for the wind, thereby inducing a strange feedback, whereby the more wood they burned the more wood they more they had to tear out of the walls to make up for the increasing cold. Eventually, the building was reduced to just structural supports and they ran out of any tinder to hand.

"Je moeder heeft een snor!" The captain yelled to no one in particular, which loosely translates as 'your mother has a moustache'.

"The sails!" Cried the first mate, and pointed through one of the gaping holes at the wooden sails sweeping past.

"Zoon!" shouted the captain, "we need a volunteer!"

Again, they all looked at Einstein, who very soon found himself dragged outside and hoisted up on their shoulders, trying to grab the sails as they whooshed past. The crew cried out encouragement every time a sail came within in his grasp, only to let out a moan as it slipped past. Eventually, he hooked a hand onto the leading edge and shot up into the air, to a general chorus of 'Hoozoon!' He pivoted in the air, watching the snowy horizon rotate at a stomach-turning speed.

"Waah!"

"Tear off the sail!" cried the captain.

"But ... I'm ... holding ... onto ... to ... it!!"

"You'll come down faster that way!"

This circus continued for sometime, with Einstein holding on for his dear life, and the crew getting more and more annoyed at the delay. Suddenly, Einstein found himself hypnotized again, like the time in Italy, only this time the clowns were a lot angrier, and it came to him that it didn't matter if he was up or down, as in both cases the amount of energy needed to break his skull if he flew off and hit the ground was the same. Finally as he again reached the highest point the wooden sail broke loose and the sail with Einstein attached flew over the heads of the Dutch sailors and landed in a drift of snow.

"Hoozoon!" They cried as they picked him out of the snow and dragged the broken wreck back to the remains of the windmill. Einstein staggered in with them, pleased to have come down, but even more pleased to still have the correct number of arms, legs and brain.

"We will have a jolly fire now," exclaimed the captain, and indeed they did for the wooden frame of the sail offered far more tinder than the remnants of the house. Soon however, as the fire burned down, they were discussed how to shore up the walls to keep out the cold, for surely they would freeze to death no matter how many of the sails they tore down.

"We could build an earth wall, from the mud and dirt," suggested Einstein; "the work would keep us warm at the same time."

This was an excellent idea, they agreed and put him to work as they crowded closer round the stove. No amount of attempting to claw his way back to his warm spot was possible as the crew insisted it was his duty.

When Albert asked how it had suddenly become a 'duty', they pointed out they could only find one shovel, and since it was his idea, he must have responsibility for it. He shrugged, and after one of the crew had thrown him a blanket to keep warm as he worked, he fell to it with the tenacity of honey badger after a plump mouse. The soil was loose and soon the air filled with snow and flying dirt, as Einstein raced to build some semblance of a retaining wall, he picked up a sweat and even began to enjoy the work, for it certainly was better than freezing to death.

As he dug into the soft bank of the dyke he soon he found the dirt had begun to turn to mud. He pointed this out to the captain and crew, and they suggested he would find it easier to build with mud than dirt, as it would stick together. The wall rapidly began to take the shape of crescent dune, and did indeed keep the wind out of the diminished house; unfortunately, Einstein had dug so deep into the dyke he had broken across the foundation soil, that hard region of gravel within the bank that forms an impenetrable barrier to surging forces of the North Sea.

The North Sea has no sense of humour and quickly expressed its innate annoyance at any impertinence of humanity by filling the fissure at Einstein's feet with freezing water. He looked down at the growing pond and pointed this out to the crew, who told him to finish off the wall and not worry about a bit of water, as they certainly weren't bothered; he dug deeper, the water filled faster and was soon up to his knees in a brackish mire. Yet he struggled on until the wall was finished and stepped out. The chill wind instantly cut through the soaking blanket. He shivered and watched with amazement as a large crevasse opened from one side of the dyke to the other, cutting him off from the windmill, the crew and more importantly from the fire.

"Um ... problem ...," he called out hesitantly. "Problem."

"Well, solve it!" They cried back unconcerned.

"Ah." He watched as the tidal surge began to eat away at the side of the trench with increasing tenacity, some strange watery leviathan tore away handfuls of soil and pushed it's way into the still waters beyond the dyke. What was but a stream, quickly became a river and this a gushing torrent, until he was forced to walk backwards from the edge of his creation.

"It seems to ... um ... have solved itself," as the North Sea disgorged itself into Holland. "Flood!"

This brought a row of heads popping up, and staring with amazement at the incipient torrent.

"Godverdomme!" was pretty much a universal cry as they scrambled over the makeshift wall to stare down at the inrushing waters, then they stared at Einstein, huddled in his blanket and holding the shovel, back to the river and once more to him. His moustache and curly hair covered in snow and mud, he smiled and shrugged as if to say, 'You did tell me to dig?'

Very soon the earthen bank began to collapse on both sides and the two parties were forced to withdraw from the encroaching the water; the windmill itself collapsed into the relentless surge.

"Stupid Tourist!" Came the distant cry as the ships crew fled away, and Einstein ran along the dyke chased by the North Sea. Three days later, during the worst flooding of Holland in three hundred years, Einstein was found tied to a barrel of herrings with only the ships mongoose for company.

Descartes at Sea

In which Descartes discovers an Ocean of Wisdom.

René Descartes had managed to escape from the purse of his landlord and retain his bed, by disguising himself as a plague victim. To bring this about, he covered his face with gobs of chewed paper and screamed relentlessly, as his manservant Pierre dressed as a priest and performed last rites by sprinkling René with rum.

"Plague! Plague!" Pierre screamed almost absentmindedly.

"Shush! Shush!" the landlord shushed back, "Are you sure it's the plague? I didn't know you were a cleric?"

"I only do last rites," Pierre sipped on the holy water, "and yes, either he has the plague or this is the worst case of constipation you'll ever see."

The landlord very quickly had René and his bed out on a cart and down to the port before anyone caught onto the possibility of the dread plague being in his tavern.

The port stank of fish, sailors and the Seventeenth century in general, which is to say it smelt about as appallingly as it is possible. To put this into perspective, take the country of France, tip it on its side, and as the all the waste and effluent pours out of Normandy it will turn the English Channel to a marvellous puce colour. Then and only could you capture the smell of a French port in the Age of the Enlightenment.

René and his manservant found themselves haggling with the ship's captain, who was a large Scotsman, a Lachlan MacBrown.

"You don't have some crumpet in there, do you?" MacBrown wondered aloud, trying to peek under the sheets of Descartes moveable bed, "and more to the point, why are you travelling around in a four poster bed?"

"I am a philosopher," René replied, "and I have dedicated my life to solving the philosophical problems of the world."

"Aye, that's grand and all," MacBrown arched an eyebrow, "but if you'll forgive me for asking twice - why are you travelling around in a bed? Why do you have clumps of paper stuck to your face, and why does your servant keep drinking that holy water?"

René tutted, "Ah, you English with all your needless questions."

Before Descartes knew it, MacBrown grabbed a pillow and buffeted him in the face with it.

"I, mister philosopher," yelled MacBrown as his beard loomed threateningly in Descartes face, "am a Scotsman and if you ever make the very stupid mistake of calling me an Englishman again … you'll wish I was."

"Ah," Descartes piped, and drew his bed sheet up to his nose.

MacBrown then roared at his crew to load the bed and it's contents into the forward storage, knowing full well that once the philosopher was on board he would be completely in his power. This was a idea that René had overlooked and found himself and his bed surrounded by barrels of French wine and bales of silk that were being transported to the Baltic. In turn, what MacBrown had failed to see, was just how much of a opportunist his passenger was going to be.

The ship was a small brigantine, which lacked a deck, one of those two masted merchant ships with lateen and square sails with a long history of landing contraband along the English coast. Smaller than a barge, it was ideal for running smuggled goods into the by-waters of the Norfolk Wash. It was fast, could be unloaded quickly and was often mistaken for a Hanseatic merchant boat. MacBrown was the sort of captain that every English naval officer hates. On the surface he seemed a law-abiding merchant who paid his dues, but beneath those waves lurked a brigand who given any opportunity would smuggle anything from rum, catholic bibles and cannons. Provided it paid well and he could get away with it. Smuggling philosophers was new to him.

Mostly MacBrown was hated by English naval officers, because he was a loud and abrasive Scotsman, who never failed to let them know this singular fact. His favourite smuggling place within the waters of The Wash, a twisting, marshy ruin of the sea, vast silted up bays filled with treacherous sandbanks and unknown tidal mud-flats, the sort of place a waterbird thrives in and people drown, and given the right amount of fog was impossible to patrol. He had been having a running feud with the revenue officer from Great Ouse in Norfolk.

His actions were set against the background of the English civil war, which had been going for about four years. A minor spat in the history of Europe about whether a king was on the same level, metaphorically speaking, as say Jesus. Some, notable those with a bit of land held the idea of kings being on the divine side. Most people, however, definitely sided with the Jesus camp and thought kings were right 'gits' who needed their heads lopped off - if only to remind them of their place in the grand scheme of things. Into this fray, smugglers were having what economists would call a period of high financial growth with low capital expenditure.

"We just have to take a detour," Captain MacBrown announced to his crew and passengers when they were barely out of the port of Le Havre in Normandy.

"How far?" Descartes asked from bed.

"England."

"You call that a detour."

"No, I call this a fist," said Captain MacBrown and held it in front of René's nose.

"Detour it is."

Even though they raced up the coast of France it was still two days before they sighted the coast of England and there they played a waiting game with the weather, hoping for a fog bank to ride in off the North Sea and cover their approach to the Great Fens of Norfolk and the smugglers den. It was during the wait that the captain sort to understand the nature of philosophy.

"So, let me get this right," MacBrown argued with Descartes as they wallowed in the English Channel. "You think that the mind, which ... well it's the sum total of what we call thought, is connected to the brain ... which is the mushy bit of the skull, by this bit of fishgig called the pinny gland?"

"Pineal gland," René' smiled, "and for someone who can strip the blubber off a walrus in under sixty seconds, you're doing rather well."

Several of the crew had joined the debate, more interested in finding an excuse to sit about than to work the deck.

"So, is this pineal gland flesh or soul?" asked Fitzpatrick, a large orange Neanderthal from the Hebrides, "That's what I'd like to know."

"Oh no, no, no, you're missing the point," enthused René, "it's not whether the soul is part of the brain or part of the mind, it's how much crumpet you can get by talking clever to the girls."

"What?" MacBrown said in amazement, "you mean philosophy is just for chatting up wenches? Have you not heard of holy institute of marriage?

"Have you not heard of prostitutes?" chipped in Fitzpatrick.

"Have you not heard of sheep?" came a crewman who hailed from the farthest outreaches of the Hebrides.

"And how much do they cost?" affirmed the philosopher, "No, trust me, better you let them think you're clever and they'll pay you just … wait, did you just say 'sheep'?"

"Arr," opined one of the hairy crewmen, "Ooh."

"Did he speak, or was that a natural movement like belching or bowels?" asked René.

"And because of this," interrupted MacBrown, "you spend your entire life in bed. You're not ill or cripple, but because that way you can spend more time … getting crumpet?"

"You have the makings of a philosopher, my obtuse friend." Smiled René. "Consider this, the purpose of philosophy is to find the right means and the best way to get through life. Yes? And so to this end, I believe that one should find a place within god's creation as close as possible from where we originated, that which we call the womb. Is not a bed like a womb, a warm and soft place, and if properly organized a place filled with the finest foods and the tastiest crumpet. All it requires is that one uses mind not to solve the riddle of the ages, but how to retire at thirty five years of age."

"If you say so," said Fitzpatrick, "Ah now, I'd never thought about it before, mind you I've never thought before. There be just one thing?"

"And what is that?"

"What's obtuse mean?"

While they had been talking the fog bank had rolled in, and suddenly from the starboard came the unmistakable report of a cannon at close range.

"Revenuers!" cried all in sundry, in much the same way, as Spanish galleons sailing through Caribbean would echo with the cry of 'pirates'.

Through the gloom the prow of a parliament's ship loomed towards them.

"In the name of the parliament of England, I call on you to heave to."

"Why did you fire on us before you called?"

"Um, we were fishing."

"With a cannon?"

"Whales!"

"Be yee roundhead or cavalier?" MacBrown yelled in annoyance.

"It's a bleedin' parliament boat! What do you bleedin' think we are?"

"For all I know, you could be whale herders."

"What in blazes does this look like?" In frustration the captain of the roundheads soldiers banged on his helmet with his sword hilt. "It's a bleedin' roundhead helmet, I just yelled out we're a parliament boat. How much bleedin' information do you bleedin' need?"

"Well, be yee friend or foe?"

"I know it's you MacBrown! I'm a bleedin' foe I expect!"

At this, a deeper fog fell upon the seas and covered them from stern to stem, leading to a chase for several hours which revolved around yelling match between the revenue officer and MacBrown, as the English boat chased them back and forth until they were lost in by waters of the Great Wash.

Meanwhile, in the prow of the boat Descartes had returned to dictating philosophy with Pierre his manservant.

"What chapter are we up to?"

Pierre who had spent almost the entire voyage regurgitating like a seagull to its chicks, collapsed on the bed.

"We're going to die, please make it quick."

"Nonsense, we're off to Sweden for a bit of crumpet, now where are we?"

"English channel, about to be either shot as smugglers or by the smugglers, or hung as being spies and smugglers. Where did you think we were?"

"The book, damn it, I've no time for discussing politics. What chapter in the book are we up to? And do we have any chicken?"

"It's floating several miles behind us."

"What's it doing there?"

"I threw it up."

"Was it off?"

"No, but I was," Pierre grimaced, "let's see, you have just proved you exist."

"I have? That was easy. What's next?"

"You need to give reasons for the laws of nature, the sun and stars, the idea of the moon being the cause of ebb and flood, on what causes gravity, the nature of light and fire, the purpose of medicine, the flow of blood through the heart and arteries. Show that all these motions are totally independent of what we think, and finally prove that the soul is independent of both God and our bodies."

"All that? May have bitten off more than I can chew. Hmm, let's attack gravity and try to knock it off before lunch."

"There is none."

"No Gravity, well that was fast. What's next?"

"No lunch, Monsieur. Gravity is still here."

"Bah! Gravity, Gravity?" René mused, "Hmm, what works like Gravity? Gravy? Something that falls down?"

"No, Monsieur, gravity is what makes things fall down."

"So you say, but who is the Monsieur here? I need an idea, any idea, I don't care which, as long as it's an idea!"

"Possibly the Earth attracts objects like a magnet, Monsieur?"

"Shush! I've got a better one - Vortexes!"

"Monsieur?"

"Don't you see," René became enthusiastic, "a vortex it's a swirling pool of water that sucks down things. It's brilliant. The word sounds just great, a vortex of water is pretty to watch, and nobody has the faintest idea what I'm talking about. You can't lose an argument if they have no idea what you're talking about."

"But won't they ask to see this great vortex which swirls about the world?"

"Ah, but it is invisible! Oh, I really am clever! The Earth has an invisible, untouchable vortex that pulls things down. They can't prove me wrong; mind you they can't prove me right. Either way, if I stitch it in a book it's going to sell. Trust me, big words and pretty pictures – that's what sells!"

Through the fog came more cannon fire, before a cannon ball shot over them passing through the sail.

"What was that?" quizzed René.

"A cannon, Monsieur, I fear the English are trying to sink us."

"They are? Time to get out of bed, no wait the English are civilised they would never shoot a man in his bed."

"No, Monsieur, but they may shoot the bed."

"You have a point, remind me not to sell you as a Moorish slave."

"Monsieur, I am a Moor."

"Alright, remind me not to sell you - period."

Another cannon ball rushed past their heads causing them to duck.

"Monsieur, perhaps we should help the English capture the ship? This after all is not our fight."

"Remind me to sell you after all - have you seen how big that Scotsman is?" René pursed his lips. "I have a better idea, tie those barrels to the end of the bed. There is more than one way to cross this river."

"The barrels are filled with salted herrings."

"Start eating."

Out of the gloom the pursing English ship reappeared. It's

prow headed straight towards them, filled with English snipers who began a desultory firing. The bullets peppered the deck and chipped holes in the rigging.

"Surrender MacBrown, and you'll only get the gallows!" came the cry of the pursing naval officer.

"Ya damned Turnip Head! What kind of a daft deal is that?" MacBrown yelled back.

"Alright, surrender and you'll only get life imprisonment."

"That's not what your mother said last night! I have a better idea!"

"What's that?"

"This!" MacBrown yelled as he turned his small brigantine into a channel he hoped was still there and then turned again. The larger frigate, it's cannons pointing uselessly in the wrong direction first ran into the channel and then ground to a halt as all too late the captain realized the Scotsman had lead him into the low tide sandbanks of the Wash. MacBrown chuckled with delight: "I'll be seeing after the high tide!"

Then his own boat ground to a halt as it also ran aground.

"Ah, yee Minky Basturt!" cried MacBrown as he and his ship's crew fell over in the brigantine's momentum.

Thus began the Battle of the Tides. The two ships stuck fast in the receding waters of the estuary, their captains uselessly slinging insults at each other as their crews fired muskets. At one point the English tried to launch a small boat to scull the remaining distance, but the smugglers punctured its hull with muskets balls and thereafter the English crew refused to leave the safety of the heavy wooden gunnels of the frigate. Intermittently fog would roll in and halt the gunfire, leaving only the fiery language to echo strangely about approaching night.

"For a Saxon, your mother's right gud in bed!"

"That's doesn't make any sense, coming as it does from a man wearing a dress!"

As night fell, the Scottish captain made his way amongst the fallen rigging and hiding crew.

"You there! Philosopher!"

Descartes poked his nose out from his pillow castle.

"Are we in Sweden yet?" asked Descartes.

"No ye daft git! You're supposed to be a clever man, is that right?"

"Are you a fan?"

"Ere yes," MacBrown placated him, "now tell me how the hell to get off this sandbank."

"Stuck are we? I was wondering why Pierre had stopped throwing up. Well, lets think. After all, that's what I do best. We are held here by gravity, pulling us down into the mud."

"Now that I could have told yee ma'self."

"You must disrupt the vortex!"

"What vortex? All I see is a sandbank."

"There is an invisible, intangible vortex that holds everything to the earth. This you must disrupt."

"You want me to blow up the sandbank?"

"Er, yes, or blow up the sandbank."

"Not at daft as I thought you'd would be. All the same, that gives me an idea."

Hidden by the mists the crew of the brigantine clambered over the gunnels hauling small kegs of gunpowder and planted them on top of the ooze about the ship. With the passing of the hours, insults fell and the tide rose. The lighter weight of the brigantine allowed it to rise first into the swirling waters and it began to move, but all too soon the faster and more powerful frigate surged forward as the wind caught its sails.

"I'll see you hanging on the gallows, MacBrown!" The English captain cried out, "Mark my words!"

"I think I'll mark your waterline instead!" MacBrown chortled gleefully as his ship eased away in the rising wind. "Fire!"

The musket men on his boat shot out the kegs they had planted during the night on the sandbank, just as the frigate came upon them. Dense plumes of smoke rose about the larger ship as the explosions close to its water line crippled it and began to list.

"Now what was that you were telling me about the gallows there?"

"Damn you MacBrown! This is the king's ship!"

"Don't you mean the parliament?"

"Gah!" was all the captain could muster in his confusion as he ordered his crew to abandon ship.

The smuggler's ship sailed on into the gathering night, as what damage had been done was right under lantern and soon they were able to approach the Suffolk cove where they hoped to unload their wares. In that time, the captain MacBrown made himself better acquainted with his passenger Descartes.

"About the bed, I've been wondering."

"Yes?" René sat in his bed reading a book and Pierre stirred a crock-pot. "Is this Sweden?"

"Well no, this is a question." MacBrown scratched his beard, "It can't be convenient and all, now then can it? Suppose ... I mean, all well and said, but a man needs to stretch his arse from time to time if you know what I mean."

"No, what do you mean?"

"Your toilet man, don't tell me you clean your arse lying about in bed?"

"That's what my man-servant is for, isn't that right, hmmm?" He looked at Pierre, who sighed heavily and went on stirring the pot.

"Ah, well, it was a question that was starting to weigh upon my mind and all."

"You must understand that the whole principle of reasoning must revolve about Doubt. It's not enough to espouse an idea of how one is going to live ones life, I must live that philosophical ideal as a principle in every aspect of my life."

"Philosopher, I'm having a bit of trouble understand even a little of what you're saying."

"Let me put it in simpler terms. I hold as a principle that one must doubt everything, before one believes in it. Yes?"

"Oh, aye, I don't doubt I'll nae see that Saxon bastard for another year."

"Not quite what I meant. You see you cannot know if that 'saxon bastard' exists, unless he fires a cannon at you, or tries to shoot you, or hang you. Etcetera, etcetera."

"Oh," smiled MacBrown, "I know he exists, he just tried to kill us. Ha, you'll have to do better than that."

"Yes, but you don't know if he exists – now."

"Of course he exists now, unless he just drowned. We just saw him, ten minutes ago. Where did he go, Egypt?"

"No, no, no, you're missing my meaning altogether, how do you know you didn't just imagine he attacked the ship. If it's a false memory, he may not exist at all. It's all about using Doubt to understand the world."

"Ya daft git, then why are all bullet holes about the ship? And what was that ruddy great explosion, just before when we blew up them powder kegs? If that don't mean we just lost that Saxon bastard, then I don't know what does."

"Yes," smiled René beatifically, "but what if you just imagined it all."

"Because, well, just because!"

"Precisely, you can't argue with it can you?"

"Of course I can't argue with your infernal logic. It's doesn't make sense!"

"And that is why I'm a philosopher and you are a Scotsman."

"Is that so?" MacBrown leant forward and tapped René on the chest. "And do you doubt I'm also the captain of this here boat? And do you doubt you are in my charge?"

"Ah."

"Aye?"

"That question is beyond doubt," René said wisely.

MacBrown chuckled, "your nae a fool," then walked off about his boat, readying the crew for either a quick landing or even a quicker retreat. One could never know if a town militia was patrolling the beach.

"Monsieur, perhaps it might be an idea to placate this captain. We may be sailing with him for another week."

"Nonsense, Michel, never explain, never explain, that is my motto. If the peasants know too much - they will revolt. Mark my words, should the rabble ever learn to read that would be the end of order, and for that matter an end to crumpet. Make a note of that, Order and Crumpet, there's a book in there somewhere."

"It's still Pierre Monsieur, and the captain isn't crumpet, even if he does wear a kilt."

Before the conversation could continue the cry of land went up from the forecastle, and the smell of salt water was mixed with the distant odour of English fields with summer haystacks, followed by the sounds Norfolk beach rushing in the distance. Soon they came to shore and ran the boat upon the white sands, which glowed in the moonlight like an Arabian desert.

Descartes poked his head up and looked over the side of the bulwark at the approaching shore.

"Is that Sweden? There's crumpet in Sweden."

"No, Monsieur," Pierre replied, "This is a dreary little country where the idea of philosophy revolves around how well they can murder the inhabitants of the neighbouring countries. They are a very practical minded people."

"Oh," said René disappointed, "you mean England."

The ship's cabin boy, a large man about forty-seven appeared before them.

"The captain says you have to get out of bed and help with the unloading."

"Does he know I'm a philosopher?"

"Aye, he said if you said 'philosopher' I was to..." but he never finished.

"Look to windward!" came the sudden cry of the ship's watch, who sat high above them in the crow's nest and pointed behind. They saw to their amazement the badly damaged revenuer's ship, still afloat and giving every indication of chase.

"Oh for gud's sake! Yee have to be kidding me," yelled MacBrown to no one in particular, but there could be no doubt about the intent of the pursing craft when they saw the flash of a musket fired. "Over the side mates, push the ship back off the beach! And throw that damned philosopher over as well!"

The crew stopped their unloading, flocked about the bed and tossed it overboard.

"I paid full fare!" was Descartes only comment, as he landed on the beach, safely beneath the covers of his bed.

The crew then set to push their boat back out to sea, but it was jammed upon the sand and refused to budge till they had lightened the load. Once more they swarmed back on deck, only to give a cry of despair when they saw the frigate had closed their escape and they were trapped on the beach.

"Everyman for himself!" MacBrown ripped back a haversack, picked up a small chest before darting forward over the prow and disappearing with his crew closely following him.

"You call yourself a ferryman!" Descartes shook his fist at the retreating Scots.

"Monsieur," Pierre asked not a little perturbed, "what are we to do?"

"Wait."

"Just wait?"

"Think, what will an English captain do when he finds a man stranded on a beach in a bed? What will he think? If he was a German captain he would leave me there, a Spanish captain would tow me out to sea, and an Italian would try to get in bed with me. The English? He will offer me a hot drink and read me bedtime stories. The English are so predictable."

"Of course, Monsieur, I have no doubt of your reasoning."

"Ah."

Soon after this the English landed and asked who he was. Upon learning he was French, a philosopher and Descartes, they stuck his head in a rabbit hole, and rammed a stick into the earth between his legs to prevent him escaping, then ran off into the night after the smugglers.

Pierre got into the bed and without a doubt slept soundly through the night.

Einstein and the War

It was a dark and possibly stormy night; when a certain patent clerk from Switzerland was accidentally inducted into the German army as a messenger pigeon.

"Left, Right, Left Right!"

As the men paraded stepped in two ridged lines, their legs flew up in the air, pointing their toes out as far possible then strove to slam them back to the ground with an ankle breaking clatter. It was excruciating to watch – unless of course you were a sergeant major in the German Army in 1916.

"Halt! To the Left - Turn!" Screamed the sergeant at arms, whose handlebar moustache was held to be a thing of beauty in his home town of Mittenwald. "All present and accounted for! Sir!" He about-faced, marched behind his commanding officer, there to pace back and forth glaring furiously at his new recruits.

The captain surveyed the soldiers, his blue eye so brilliantly glittering it seemed unreal and in some way made up for the missing one. He had lost that one while sparing with bayonets to his nanny. The finest of Prussian officer Junkers class and had the scars to prove it, having attended, under the wise tutelage of his father, and his father's general, the Studentenverbindung, a German student fraternity where delinquent boys and wayward children learned how to fence with the sole aim of losing parts of their face. Indeed, any child, returning home on midterm break, without something approaching a puncture or a slash on his face was considered a disgrace to the entire Prussian upper class, and would have been fostered off to learn goat molesting in Slovenia. His own particular fencing scar was a vivid lightning bolt, much admired by his fellow officers, so much so he considered getting a second one, but demurred on consideration he couldn't afford to lose both eyes.

"It has come to my attention that we are harbouring a conscientious objector." He slapped his leg with his riding crop and silently winced.

The sergeant at arms snapped open his holster and with his pistol charged forward pointing it at the men. "Who sir? Let me shoot him!! I have dum-dum bullets!"

"Not yet, Sergeant, that is not the German way, after all, first we must break his spirit then we go to the target practise."

"Yes sir!" and the sergeant randomly shot a pigeon out of the sky.

Days before all this our hero had enlisted.

In the middle of the year Nineteen Hundred and Sixteen, Einstein had made the mistake of visiting his parents in Berlin. It was a dark city in the depths of the war, it's munitions factories were exploding with wealth, coffee was being ground from acorns, and groups of women handed out white feathers to any man with the right number of limbs; or if they'd been gassed - enough of his eyesight left to avoid being run over by Berlin trams. As Einstein walked back from the train station, he become aware of the sombreness' of the people, they had become a melancholic people far from the Belle Époque, their smiles faded quickly on waking from pleasant dreams of the night to mornings of madness filled with Imperial nightmares. All of this contrasted, with his own cheerful disposition, and as he stopped in front of a tavern, he wondered if he should drop for a quick ale before he continuing home - when from across the street a shrill Fury pointed at him and called out:

"You there!" said a woman with a face so lined you could have used it for to frighten bears.

"Me here?" he replied.

"Yes! You there! Why are you here?"

"I'm over here."

"Not over there, what are you doing here?"

"You're over there, I'm over here!"

"No, you're not! You're still here!"

"Your mistaken, I assure you, you're over there, I'm over here, this is a geographical fact, so I'm lead to believe."

"Don't talk twaddle with me boy! Why aren't you at the Front!"

Einstein looked behind himself at the front of the tavern.

"I am at the front."

This was too much for the woman and promptly brought traffic to a halt as she strode majestically across the street. To the amusement of the tram drivers, who were sick of running down blind veterans. Planting her feet in front of Einstein, she towered above him an oak of a grandmother, her bustle and feather hat making her seem some antediluvian creature left out of Noah's census.

"Coward! You are not at the front! You are a traitor to the Empire! To the Kaiser! To the free world!"

"But we're not free, are we? I mean we maybe part of a representative democracy and all, but ..."

She slapped him, pulled a large peacock feather, which had been died white from her hat and handed it to him, before once more launching herself across the street, seizing up the Berlin trams schedules.

"You should not be here!" She cried over her shoulder and disappeared into a milliner.

"I'm not, you ... ach ... never mind."

Absent mindedly he stuck the feather in his breast pocket and wandered down the street wondering what all the fuss was about. He had spent the last four years stamping patents in Zurich, and had no idea that the civilized world was not so civilized as it like to think itself, as it was at that very moment tearing itself apart and blowing to bits hundreds of thousands of young men on the muddy fields of the Somme.

"Halt!" From across the street came another war scream from a woman who had never seen a wound worse than a braised knee. "Stand and be accounted for!"

Einstein held his finger up.

A herd of wildebeests dressed in crinoline and albino peacock feathers trampled their way across the street, putting the Berlin tram schedule so far behind that half the drivers stopped to have a smoke and enjoy the day's entertainment. The Berlin's Women Auxiliary Defence and Polo Club was in full war regale, thirty infuriated women stood around Einstein glaring at him, and if they had had but one polo mallet amongst them he would had been the afternoon news.

"Stand and be accounted for!"

Einstein held up his finger again. "One."

This brought an infuriated shriek from the head wildebeest. "Mock me not, sir! We are here to ask you where you stand? Out with it! Where do you stand?"

Einstein looked down at his feet, he eyes rose back up to the angry buffalo and then back down to his feet.

"Here?"

"You wretch! We of the Berlin's Women Auxiliary Defence and Polo Club are not to me scoffed at - we have husbands you know! Husbands in high places! Husbands who wield immense power and we wield them! Now answer our question! Do you stand with us or against us?"

He judged the distance to them to be sufficiently close to warrant a 'with them', and this was his answer.

"Here with you."

"Then why sir, are you here?" The unflappable head buffalo continued.

"This 'is' where I'm standing? It's all relative you know, if I was over there, then I wouldn't be here, I'd be somewhere else."

She blinked several times, and then turned around to discuss this latest finding with her ambiguous drove. There was so much hyperbole and praises to the Kaiser that he started to edge his way further down the street, wishing he had a bicycle to escape on.

"Halt! You must be accounted for!"

Einstein sighed.

Several hours later he found himself at the recruitment office pursued by a regiment of feather waving matrons intent on his joining the battle for the Fatherland.

"Name."

"Einstein. Albert."

"Occupation?"

"Um, is there a back door?"

"We only have a door to the Front."

"That's what I was afraid of. You see, I'm not supposed to be here, or there, ach ... all these different places I'm not in."

The Lieutenant was hypnotised by this outburst for a moment, then pointed behind him. "We'll have a psychiatrist look you over, go that way, second door on the right. Look for a lunatic."

As Einstein wandered off, he heard behind him the ruffling of feathers and cries of disappointed wildebeest, as the Lieutenant yelled at his junior - "Sergeant, take these women to the train station and send them to the Eastern Front ... for whore duty."

He opened the door and stepped into a smog filled room. At once an obscure shard of a Mesopotamian figurine appeared in one hand and a lit cigar in the other, as he was pushed from behind to a leather divan, which he dimly saw was covered with an ornate Persian rug.

"We will continue from the last consultation. Tell me about your dreams?"

In the gloom it was hard to see, an aged man concealed himself in the shadows, his eyes sunken and his bearded speckled with grey. He sucked his teeth and chain-smoked cigars at a prolific rate, so much so the room was filled with a pea soup of pungent smoke. Einstein joined in the festival of smoke and lay on the couch; staring in puzzlement at the Sumerian statuette and trying to remember where he had meet this psychiatrist.

"Err ... have we meet?"

"You have amnesia? Why didn't you remember to tell me this? Oh, yes, of course. You have amnesia. I am Sigmund Freud; you came to me at 14 Alexander Strasse, Vienna, Eleven o'clock in the morning in the year Eighteen Hundred and Ninety-Seven. You consumed the varnish off the piano and attempted conjugal relations with my wife. I diagnosed you as suffering from acute ergastic hysteria of the lower intestine. Have you no recollection at all?"

"Oh, that Sigmund Freud. Do you have any more of those ... um ... delicious buns?"

"The cocaine ones? No sorry. They're poisonous, don't you know? Now then, let us continue our dissection."

"Don't you mean - discussion?"

"You have grown up and learned larger words I see. Here have this," he tossed another Babylonian vase in his lap. "Now then, we must journey into the heart of your soul, the very apex of your being, to a place where ..."

"My cigar has gone out."

A box of cigars, followed by a box of matches landed in his

lap.

"... to a place deep within your psyche ..."

"Ashtray?"

An ornate and intricately carved ebony dish from Greece joined the growing collection.

"Now please, no more requests, we have a war to win you know."

"That's the problem, you see."

"Do you object to going to the war," said Freud as he made a note of this objection.

"Well, yes, I'm no longer a German, I lost my passport and can't get it back."

"Ach yes, I see, you have repressed your need for conflict and wish to hide within your Super-Ego having rejected the identity of your Fatherland. This is a common complaint I am seeing often amongst young men of your age."

"No, I lost my passport, and I can't get it back, and I'm not that young." Absent mindedly he dropped the match, missing the ashtray and falling instead onto the Persian carpet, which smouldered for a moment, there broke into flame.

"Have you ever subscribed to Girly magazines?"

"What no ..." Albert jumped seeing the smoke billowing from beside him. "Fire!"

"Ach, yes, you smoulder with desire for the female form, adolescents often express the need for the female form by pyromania."

"I'm Thirty-Seven! And No! I don't smoulder for the female form - I mean Fire!"

Freud gave him an antique fire extinguisher and they fled the room. Later Einstein was court martialled for the wilful destruction of a military establishment and sentenced to be sent to the Western

Front. In his defence, he stated he was still wasn't in the Wehrmacht as he had lost his German citizenship years ago, and it was all a terrible mistake.

For this, they posted him to both the Western and the Eastern Front.

The captain wished he were back home on the Front Line evading bullets and releasing gas at the English. The shot pigeon landed in front of him, he kicked it with his toe to one side, and focused his monocular attention on the conscripts. They were as average a set of humans as one could expect, they shuffled and listened nervously to the distant tempest of the Somme, where a human storm of blood and terror was heard as dull thunder, the thumping detonations of high explosive was destroying life and military objectives. Yet, here on the parade ground, the only psychosis was in the unblinking stare of the sergeant major as he paced incessantly behind the captain.

"We will have no cowards in this platoon," the captain continued, "the Kaiser has ordered you to give your all for the Fatherland. This is not an idle command and the Kaiser is never wrong, we will be victorious, even if we all have to die in the process. I want you all to know, that the Imperial German Army has never lost a battle, that ... Yes you there, what is it?"

He pointed at a spectacled history teacher, who had held up his hand in the front line.

"Sir, what about the battle of Jena against Napoleon?"

"That was a tactical withdrawal," posited the captain, "the glorious Imperial Army prevailed with little loss of life and a great victory ... oh now what?"

"Sir, we lost forty thousand men,"

"They are missing in action. They will turn up one day, and

when they do, they shall be shot for desertion. You have my word on that."

"But sir, the battle of Jena happened one hundred years ago. Even if they did desert, and that's ridiculous, they'd all dead by now anyway. Besides, the French army conquered Prussia."

"Sergeant, shoot that man."

Bang.

"We will have neither cowards nor historical critics in this platoon. Sergeant, march him to the Front. The Kaiser demands a victory and the Empire demands a sacrifice."

"He shot me!" Said the wounded history teacher.

"It's for the Kaiser. You will be awarded a medal if you die."

The man fainted and was carried off.

"Now then," continued the captain unperturbed, "as I was saying, it has come to my attention that we are harbouring a conscientious objector in our midst, and I assure you, this came as such a novelty to me I had to look it up in the military handbook. I would not have believed that such a loathsome reptile as a conscientious objector could have existed within the ranks of the glorious German army. Einstein step forward."

Albert hesitantly motioned a toe.

"Look there, everyone, behold the traitor. No! Sergeant, you can only shoot one of them a day, it's in the regulations. I want you all to study this fiend, note his non-regulation moustache, a sure sign of a conscientious objector. If it weren't for the likes of him, we would have won this war, back in Eighteen Hundred and Ninety Five."

No one dared to contradict him.

"Now then, Sergeant, tell me, what is the most dangerous, most deadly, most loathsome duty we can assign this fiend to?"

"Fighting in the trenches, sir!" The Sergeant saluted and hit himself in the temple with his pistol.

"That's it?" The captain stared at his number one, and slapping his riding crop against his leg. "That's the best you can come up with? We all go to the frontline. No, I want something even worse than that. Something really bad."

"Latrine duty, sir!"

"Will that get him killed?"

"He might drown."

"No, no, I want a sure fire method. Think man, think, what's the best way for him to attract enemy bullets?'

"Um ... messenger runner sir!"

"And why is that dangerous?"

"We could get him to run messages to the enemy, sir!"

"Brilliant! Remind me to promote you to corporal one day."

"Sir. I'm already a sergeant!"

"Yes, yes, but you have to work your way up through the ranks don't you. None of this 'I'm a captain because my grand papa is a general nonsense'. Now then, Einstein, you are now the divisional runner. Anything to say? Hmm?"

"Sir. I lost my passport, I'm not really a German anymore."

"We shoot deserters, you know?"

"Nothing to say, sir!" Einstein saluted and clicked his heels.

"Much better. Right then, corporal, to the Front!"

"I'm still a sergeant - Sir!"

"Yes, yes. Now off with the lot of you and don't come back until you've captured Paris. We will give those English dogs a muddy grave."

There was a continuous high pitched shriek in Einstein's ears, all he could see was brown, it didn't make sense, little made sense out here, in the wildness of the No Man's Land. After a while, the ringing subsided to be replaced by a dull thumping, he could feel the ground vibrating from the impact of English high explosive, the brown turned out to be the mud in front of his face and he discovered he was lying on the ground, his hands upon his head and screaming in fear.

"Wake up, you're needed."

Einstein woke up and discovered instead he was in a bunker near the German command for that section. Far from being in immediate danger he was tucked safely under thirty tonnes of concrete and sandbags, he lifted his head from the pillow and fell onto the floor, having forgotten again he slept on the top bunk.

"Come on sleepy head, war to win and all."

Albert attempted to crawl back into sleep, but the soldier occupying the bottom bunk steadfastly refused to yield his blanket, buffeting Einstein with a pillow. "Get orf!"

He staggered to the door, dropping his helmet on his head and wiping the sleep from his eyes. Contrary to his expectations, and those of the Sergeant-Major-Corporal who had recommend him for the job, the divisional runner had turned out to be the safest position on the entire Front. For the efficiency of the German war machine had meant there were so many lines of communications from the divisional headquarters to the front line and the reserves, that they almost never called upon him to do anything else than make very bad ersatz coffee or run messages to the French mistresses of the German officers, far behind the dreaded Front-line.

"You there, take this note and go down to the French chef. Tell him we want quail eggs for lunch and not pigeons."

Einstein staggered out into the morning; about him a brilliant sunlight penetrated the smoke, the stench of dead donkeys and delicious cooking from across the compound.

"Quail? Mon Dieu!" Monsieur A. Boulanger stirred furiously." Where do I find quail from, here? There's a war on you know? Has he gone mad? Pah! Answer me not, they are all mad,"

The cook was an irate Frenchman, who had been captured in the early days of the war and ended up with the German high command, once they had discovered he was the head chief for Maxim's in Paris. Since the Geneva convention hadn't cover the art of French cuisine in modern warfare, he was swiftly set to task to reproduce some of the finest recipes of the Belle Époque for Fritz Wilhelm Theodore Karl von Below, general of the Imperial German Second Army. This title made no impression upon the Frenchman who consistently sabotaged the meals with whatever came to hand, and it is in a little known episode in culinary history that Monsieur A. Boulanger is a credit to his nation and the world of gastronome.

"Well?" He stared angrily at Einstein, "what do you want?"

"Breakfast?"

"Not a chance, here have this bowl and go find me some quail eggs."

Einstein held up the bowl, his eyes drifted dreamily to the sideboard filled with fresh white loaves of bread and oranges from Africa.

"Quail eggs," Boulanger pointed with his spoon, "try the forest."

As Einstein shuffled off, he saw Boulanger muttering to himself in French, and pouring sump oil into the General's pancake mixture.

"Wait, take that bicycle, it will be quicker."

Albert's eyes lit up like a parachute flares just before a mortar bombardment.

He rode the bicycle erratically amongst the shell holes, expertly weaving his way amidst the morass and barbwire, for whatever else was said about Albert Einstein he was a genius at riding a bike. After an hour of meandering he finally discovered his forest. He also saw another German solider waving to him from far to the left, he waved back comradely, never thinking that the solider might be warning him of some immediate danger and so rode unperturbed into the darkening wood. Tall oaks and pine trees royally ascended about him and he had the feeling of a certain undeniable magic that one often finds in the midst of a sudden still, once one has quit the bustling city. He stopped to admire the soaring foliage, spears of light fell into the gloom opening the forest with solidness and strength, and it was if a giant had thrust in a lance of brilliant illumination.

It was at this moment he realised he was lost.

Staring about, he saw no path leading out, nor and more importantly no quails or their eggs.

"Gott! Gott! Gott!" He muttered and dismounted the bicycle to search for a road, as he walked along he heard the undeniably sound of soft crackling of feet in the gloom behind him. He froze and so did the steps. He stole another step and leaves mimicked him. Turning around, he stared at the spot he imagined his stalker was hiding. "Hello?"

There was a strange cough, "gg- gollum." The leaves parted and a pallid face framed by the curling moustache of a lance-corporal stared back at him.

"Can I borrow your bicycle?" It was none other than a certain Adolf Schicklgrüber, who also was a messenger runner for German Army. The path of the two messengers had crossed many years ago; although they were unable recognize each other, hidden beneath their respective moustaches.

"I must find quail eggs for the general," replied Einstein, "do you want to tell the general I could not get quail eggs because you were too busy riding my bicycle?"

"I have something better than quail eggs!" Riposted Schicklgrüber, he coughed again, "- gollum."

"Quail?"

"Even better! I have a prisoner!"

"Ach, you can't eat those you know."

"That would depend on how you cooked them."

Einstein looked shocked, but Schicklgrüber recovered the moment by laughing it off, which was choked off by the strange cough, "Gollum."

"Are you alright, that's a wicked cough."

"Trench fever, I'm getting over it." He hacked up yellow bile and spat into the undergrowth. "Gollum! - Do you want to see him?"

"Has he got any eggs? If I don't show up without any eggs, I'm going to have one really annoyed General Fritz von Below."

"Phah! Officers! I could show them how to run a war. Come on, help me take him back, we'll get the Iron Cross for it." Schicklgrüber persisted.

"You know, I'm not really part of this war. But very well, lets see your prisoner. Maybe he can show us the way out of here."

"Gollum!" Spitting green phlegm upon the ground.

"You really have to get that cough seen to."

They made their way through the forest, with Einstein carrying the precious bicycle over his head, partly for it was easy, and partly under the desperate hope that it might stop a stray bullet - far from the madding front though they were. When they came upon him, J. R. R. Tolkien was lying on his back staring up at the trees, thinking it would be a nice place not to die in. He had been suffering from trench fever and had been lost from his platoon for days now; he dearly wished someone would give him a hot cup of tea, or at the very least a hot cup of whiskey.

"I grabbed him in the night, I was slinking over the mud

trying to find a juicy rabbit to eat, and there he was, gollum!"

"Stop making that sound!" exclaimed Einstein, "it's really, really strange."

"Hello chaps, oh, Germans I see. Wie Gehts?" Tolkien asked in perfect German.

The two Germans looked at each other in surprise.

"You are not so different to us then, are you?" said Albert.

"My grandparents probably knew your grandparents, in fact." Tolkien smiled wanly, "I say do either of you chaps have any whiskey?"

"Gollum." Coughed Schicklgrüber.

"Nasty cough that, got a bit of the trench fever, have you?" Tolkein asked politely.

"Silence swine! You are my prisoner! Gollum."

"Steady on old man, I do wish you'd stop all that threatening." He looked at Einstein, "He's been like that for hours now. Staring at me from the bushes yelling insults and spitting. Dreadful sense of humour."

"Why are you lying in the leaves?" Einstein asked, "Are you wounded?"

"Touch of the dreaded lurgy. Had it for days now, I've given up running to the toilet. Now I just roll."

"I thought you said you had captured him?" Albert quizzed Adolf.

"He lies!" Schicklgrüber gesticulated, "I mean ... yes I did. But he got away!"

"By rolling?"

"Clever swine!" Shrieked the Schicklgrüber. "Gollum!"

"Yes, quite."

"Well, you hold the bicycle, I'll carry him. We can't be too far from the Front. Or are we at the Back?"

It had begun to rain, a sombre unrelenting downpour that in the grey north of the Shetland Islands farmers would have considered necessary for a fine day. The dreary gloom was relented only by the occasional incendiary sparkle of Very flares burning green, red and white phosphorus gaily as they descended on parachutes, and the distant screams of donkeys drowning in the mud. Through this, trudged Einstein, Tolkein and Hitler, surrounded by grey mists which swelled up from the shell holes, swirling about them. It was a dismal, unnerving scene, filled with the distant rumble shattering explosions; each nightmarish scene would have given Lady MacBeth the screaming willies. They had been lost for several hours, wandering back and forth across the lines, alternately shot at by the Germans for one of them having a British uniform, and by the British for riding the bicycle on the wrong side of the shell holes.

"Gott in Himmel!" Screamed out Tolkien, "I'm British!"

"Who are those two with you?" Came the muffled voice of a Lancashire fusilier. "Punch and Judy?"

"Die you English dogs!" Screamed Adolf. "Gollum!"

This brought a rattle of machine gun fire as the trio dove into another muddy hole.

"Not bloody likely! And get that cough seen to matey!" Laughed the fusilier.

They found themselves in the bottom of a shell holes filled with mostly broken barrels of wine.

"Look chaps," propositioned J. R. R., "it's been awful nice of you to carry me all this way. But if you could see your way to letting me go here, I'll see if I can get you sent to a nice prison farm down in Sussex, or even the Isle of Wight. They have lovely weather this time of year and even fishing jaunts."

"Silence you ... we have a secret weapon! Gollum!" Adolf

spluttered.

"We do?" Einstein looked puzzled.

Hitler pointed at the bicycle.

"That's not very secret you know," pointed out Tolkien, "I have one of those back in Oxford. Although, mine doesn't have red tassels on the handlebars. Hang on chaps, it just occurred to me, I'm a lieutenant."

"Yes!" Grinned Adolf, "A lieutenant, we shall get two Iron Crosses for you!"

"Yes, but, I outrank you both."

"What?"

"Well, I'm the senior officer here, I'm the one who should be giving the orders."

"That' doesn't make sense." Screamed Adolf. "You're the enemy. You perverse English son of a kipper. I captured you!"

"That doesn't preclude the fact, that you must obey a senior officer at all times, now then does it? The whole basis of chain of command revolves about this singular point."

Adolf became apoplectically blue in the face and his hands seemed to claw the air in frustration. Yet he could not fault the reasoning.

"But ... but ... but!"

"He may have a point you know." Einstein lit his pipe and considered the possibilities of being captured by his prisoner. "He could even order us to surrender to the British. We'd probably end up having to wait out the war, over in England. I expect."

Hitler fell over and began writing about in the mud. "Gollum, gollum, gollum!"

"Nasty cough that."

"Swine! I will no longer take these insults to German pride!"

"I wasn't having a go at Germany, just you old man." Tolkien pointed out with his pipe as he hung over Einstein's back.

"You are both traitors to the glorious German Reich!"

"I still don't have a passport you know."

"Enough!" Screeched Adolf and pulled out a British hand grenade he had taken from Tolkien. "I am in charge now! I give the orders! I have a secret weapon!"

"Careful old man, those things are liable to go bang you know. Besides, that's one of ours. You do know how they work, don't you old boy?"

"Now you die!" Adolf screamed and dropped the pin in the mud, the same mud that had congealed about their feet and held them fast.

"Oh, bother," noted Tolkien and rolled off Einstein's back. "That's torn the Macintosh. Throw the bomb!"

"I'm stuck!" shrieked Adolf. "Help!"

"Grab the ring! The ring!" Yelled Albert as he dove forward.

"Never mind the ring! Just throw the bomb!"

"The ring is mine!" Shrieked Adolf and shoved his hand in the mud to grab it back. "Gollum!"

They rolled about in the mud, fighting over the grenade's safety ring. All the while Adolf held on to the grenade, having forgotten it's original purpose. Until, with a superhuman effort he ripped the ring out of Einstein's fingers and fell back, freeing himself from the mud, before running over the lip of the shell holes screaming - "It's mine! The ring is mine!"

He disappeared and there was a short crump of explosion, followed by an angry shriek - "You swine's! I've have been exploded!"

Einstein carried the sick Tolkien across the lines of the battlefield, before they were captured by a group of drunken New

Zealanders. He was then give the King George Cross for Bravery before being charged with spying, and with desertion from the Germany army. He protested that he still hadn't got his passport back, and so after a quiet word with his commanding officers, Tolkien had Albert released and sent on his way back to Zurich.

The bicycle was never seen again.

Newton and the Case of the Apothecary

Even Newton had enemies, most of England.

Gottfried Wilhelm Leibniz and Isaac Newton were the bitterest of rivals. Leibniz had once tried to assassinate Newton, from paid hoodlums throwing knives in the dark, to shepherds driving burning sheep through Newton's lectures. For his own part Newton had sent exploding clocks to Leibniz's house, once had him excommunicated by the Sister's of Divine Integument. It was a rivalry that ate at the heart of modern science. Newton never stopped from blaming Leibniz for everything that happened to him. Leibniz in turn would often yell out Newton's name whenever something went wrong. The two giants of science really had it in for each other.

Newton stared at the concoction that his manservant Hedby called tea. It seemed more akin to pond scum mixed with bracken and left to rot in a foetid cesspool.

"Hedby! Are you sure you read the label?" he yelled out down the hallway.

"Yes sir!" came back the distant reply. "It was in foreign!"

"And you can read foreign?"

"Yes sir!"

"What did it say?"

"I have no idea sir, it's all Indian to me."

"But you just said you read foreign, damn your eyes, how did you read it?"

"As King's English, milord."

Newton forehead suffered a minor geological realignment trying to understand this logic.

"So what did that say?"

"Just add hot water."

"That doesn't make any sense at all."

"Just following instructions, milord."

Newton carried his cup back to his table and stared at it. He was worried if he poked it with a spoon a toad might jump out. He sniffed it, found it atrocious and once more he yelled down the hallway.

"Hedby, are you sure this is tea?"

"It is foreign, milord."

"Bring me the packet."

The packet turned out to be a small well made mahogany case covered in alchemical symbols and Latin inscriptions with the words 'Proudly made in Britain, Mulbury and Sons Esquire, London' stamped in gold leaves on the bottom.

"Hmm, well it is British," Newton mused and sipped the tea.

An hour later he was naked and running around the Great Court chasing butterflies. Afterwards he had been caught by the university constables as he tried to climb the clock tower, and he had only escaped expulsion from the university by explaining he was performing an experiment in both gravity and hypothermia.

The next day Newton sat in the corner of his room wrapped in a sheet and sweated out the poison.

"Damn your eyes!" he railed at his manservant "I could have been poisoned! Now for the love of Zeus, where did you get your fix?"

"Milord?"

"That tea, damn you, where did you buy it?"

Hedby picked up the box and looked at the label.

"Would it be London? Milord."

"Hedby!" sweat poured into Newton's eyes, "tell me now, and

damn the shepherd who fathered you, I'm too busy dying to argue with you."

Hedby picked up the box and stared at the ceiling as if pondering some profound problem. "Mulbury and Sons Esquire, you know milord, I don't think I've ever heard of them."

"But you must have bought it somewhere, you blithering baboon!"

"Ah, no milord, I do believe it was delivered here."

Newton looked as if a bolt of lightening had hit him.

"You mean?"

"Yes, milord."

"Someone miscreant has deliberately sent me poison?"

"Oh no," Hedby smiled, "I was thinking you had accidentally been sent poison, but I suppose the other is true too, milord."

"Leibniz!" Newton used Leibniz's name like a profanity.

Hedby rolled his eyes. "Yes, milord, I expect Leibniz has tried to assassinate you, yet again."

Newton sat up and dragged the sheet across his face to remove the sweat.

"Damn his eyes! He never forgave me for discovering gravity!"

"Yes milord, I expect Jesus thinks the same thing."

Newton stood up with a ferocious expression on his face, as the sheet fell to the ground.

"Hedby!" he pointed a finger to the roof, "I must investigate this, we shall get to the bottom of this or my name is not Newton!"

"Yes milord," Hedby smiled, "shall you be needing pants today?"

"Waah!"

The next day Newton began his investigation - mostly this

involved him yelling at people till he found a result that suited him. It was only after several hours that Hedby pointed out that the address on the box was to the university apothecary Wilhelm Schmuck.

"Schmuck? This name is familiar, I'm sure I've come across his name in a text somewhere. Interesting."

"Would milord require a cup of tea?"

Newton was distracted with his thoughts and opened a dusty tome on the mantelpiece. "What? Yes. Ah, here it is."

Hedby walked out of the room with the small box, made another cup of tea and three hours later Newton was up the clock tower again yammering at the clouds.

On the third day Newton was confined to his bed, sweating even more. He held his hand to his head as if to block out the hammering headache. Hedby stood by the door holding a cup of tea.

"Hedby," Newton sighed, "is that what I think it is?"

"Why yes, milord. Your morning cup of tea."

"And curse your eyes, is that not the same type of tea which you did feed me yesterday, and so the day before."

Hedby grinned. "Yes, milord. As you did ask for."

Newton's eyelids fell and he sighed.

"I think I shall skip tea this morning. Make me some chocolate drink instead, and Hedby…"

"Yes, milord."

"Stick your head in the fireplace."

"Yes, milord."

That afternoon Newton dragged himself out of bed and into his study, there he discovered Hedby, his head covered with soot, polishing the pewter. Outside streams of addled students flowed past his window, merging with the general concourse of unconsciousness that was the university of Cambridge.

"Hedby," Newton groaned, "I order you not to make me anymore tea unless I explicitly command you. Is that understood?"

"Yes, milord."

"Now then to this problem of my poisoner." Newton gave Hedby a hard stare "Assuming that it is not you, and Hedby, just out of interest, having you been trying to poison me?"

"No, milord." His face covered black in soot. "Not unless you explicitly command me to."

"Hmm." Newton shook his head, "Anyway, let me consider this box."

He held it up to the light. It was a cube in shape, and covered with obscure alchemical script that even Newton with his vast expertise in the field had trouble deciphering.

"And you say this was posted here. Ah, yes, Schmuck esquire, care of Cambridge University - Apothecary to the King. Well then, let us to master Schmuck." He said as he picked up the dusty tome again and began reading. "It says, an apothecary at Grantham. I am surprised to find he is here. He worked on a spell to craft Homunculus via the process of preformation of the soul. Hmm."

"Yes milord," Hedby smiled.

"Nothing." Newton shook his head with annoyance. "Now then, I want you to take a letter to master Schmuck, what college is he at?"

"I believe he has a shop in the village, milord."

"A shop?" as if the novelty of this possibility had never entered him mind. "You mean he is a tradesman?"

"Yes, milord, hence the title Apothecary to the King."

"Do we still have a king, I thought we had that nasty fellow Cromwell?"

"Dead, milord."

"He is? When?"

"About thirty years ago, milord."

"Damn you eyes, why didn't you tell me?"

"I hadn't been born then, milord."

"No excuse, here wait while I write this letter."

His sheet still draped around him, Newton seemed more of a Greek philosopher abstracting a geometric principle than an irate Cambridge don. The sound of quill scratching across parchment filled the room for many minutes, as Hedby stood there with blackened face and hands full of pewter. Newton's fingers became covered with ink. It went on for so long, and so involved that Newton dehydrated from his convulsive sweating he absent mindedly asked for a drink.

He was finally cornered as he started to sprint down Trinity Lane stark naked, and on the fourth day, the university porters had tied him to his bed, and refused to release him, until he promised to keep off the tea.

"Hedby," he whispered as he lay on his cot, "come here Hedby."

"Yes milord?" Hedby edged his way forward.

"Did I not command you, that you must not make me anymore tea?"

"Yes, milord, you said I must not make you anymore tea unless you explicitly command me."

"Hedby, you blithering son of an incontinent toady, when I get out of here I'm going to tie your eyes to college flag pole and run them up the flag."

"Thank you milord."

It was only a week later that Newton finally made it to the

shop of Wilhelm Schmuck the Apothecary. The shop smelled of mystery and garlic. Newton hated garlic but he loved mystery. He stood there peering into the shadows. It was a large room filled with filled endless racks of bottles that reached up to the ceiling, bats and dead squirrels hung on ribbons from thick oaken beams, upon which enigmatic alchemical scrawls in Greek, Arabic, and Welsh proudly declaring the specials of the week:

"Pancreas of Walrus: 2/10d"
"Soul of Leek: 9/"
"Vitriol of Duck /1d"

Upon the counter were various scales, weights and ledgers; these in turn also were covered with diagrams, cryptic Latin and Welsh; and most extraordinarily, in fact it was the most mysterious aspect of the shop, was a six-foot moose standing on the counter eating cabbages.

Behind the counter half hidden in the gloom, was a middling tall man with thick spectacles and an appallingly bad wig. He appeared to be preoccupied with writing in one of his journals of accounts, the scratching of his pen made him seem some strange crepuscular mammal, like a dormouse or a rabbit that only ventured out of its cosy nest when the sun was setting and the moon had yet to rise. Wilhelm Schmuck was in fact Gottfried Wilhelm Leibniz, who had disguised himself in yet another of his brilliant plans to bring down Newton.

Newton, on the other hand was simply too amazed to find a moose in the middle of Cambridge to worry who was behind the counter, and found he had no choice but to ask why the moose was there.

"Why do you have a moose upon the counter?"

"Sven is my familiar," Leibniz smirked seeing Newton had not recognized him.

"A moose?"

"They don't have all to be owls and toads you know."

"I thought moose were extinct in Great Britain."

"I imported it from Belgium, moose eggs are very rare."

"Moose eggs? Moose lays eggs? Show me."

"I said they were rare."

"Are you sure you are a Apothecary?"

"I have a sign on the door."

"Yes, but that doesn't necessarily mean you're a real Apothecary."

"Yes, it does. Want to buy a moose egg?"

"You just said you were out of stock."

"Sven has been broody for about six months, any day now."

"Isn't Sven a male moose?"

This brought an uncomfortable pause from Leibniz.

"Yes."

"So?"

"It is a Belgium moose."

"Ah." Newton at this point gave up the problem.

"Would you like a cup of tea?" Leibniz von Schmuck leered, and the hour's later Newton was found up the clock tower with the moose.

Einstein and Buffalo Bill

Europe was once the Wild West – A. Einstein's journal

A tall man had entered the Einstein shop, dressed completely in black with a dark felt ten-gallon hat. While he seemed to radiate menace, his trim moustache and white goatee beard gave lie to any sense of peril, for with a twinkle in his eyes and a constant grin he seemed more a carpetbagger selling snake oil than a rough riding cowboy. William Frederick Cody, more famously known as Buffalo Bill, had gimlets for eyes, he didn't know what gimlets were, but he was darn sure he was scary looking. In his hand he held a copy of a dime novel which had been written about his exploits from Bad Lands of Dakota to his days as an U.S. Army scout and Pony Express rider, in it he was amazed to find out that he had killed over one hundred Indians and even more astounded to discover he was the husband of over two thousand squaws. He scratched his Mississippi beard and stared down at the novella. Its cover was a poorly rendered facsimile of himself dressed as a rabbi chasing Cossacks out of Warsaw, it was this that confused him the most, for not only had he never even been in Warsaw, he didn't know what 'in tarn'ation' was a Rabbi.

Albert who was helping out his uncle, Rabbi Silverstein, in his shop came out to the front desk on hearing the door chime, he took one look at the foreboding stranger and he knew in an instant that this was a genuine sewed into his chaps, burned by the sun, trampled by Russian Cossacks cowboy. At the age of thirteen Einstein, like all boys, fascinated by the wild west of America, or at least the version that Rabbi Silverstein had written about in his cheap shekel novels.

The gimlet like eyes of Buffalo Bill stared down at the youthful Einstein. Albert was amazed, for the eyes really were like gimlets, just as Rabbi Silverstein had described them. Bill solemnly held up the book in front of Einstein face.

"I'm looking for the varmint that tha'r darn gone wrote this."

The adolescent Albert gushed, his eyes fell open as if Annie Oakley had shot them out.

"You're ... you're ... you're."

"Looking fur a varmint."

"You're..."

Bill sighed. It was one thing to be famous; it was another to be so famous people couldn't talk to you.

"Boy! Do you know who wrote this?"

"You're..."

"Hello sir, can I help you." Einstein the elder from the storeroom and looked Buffalo Bill up and down.

"I was told that this here shop is selling these books about me." The gimlet eyes bored holes in Silverstein's head.

"But... but ... but..." Silverstein attempted, his powers of speech frozen by the menacing stare of the intruder.

"You're... you're..." Albert continued to babble.

This was too much for the cowboy and he pulled out a revolver that had killed more Indians then he had ever meet, at least according to Rabbi Silverstein's account, and shot out a gas lamp.

"I'm looking for some answers, and I intend to get 'em!" Buffalo Bill thundered.

To his great disappointment, however, the two shopkeepers just stared at the shattered gas lamp and then back at the American cowboy with his extraordinary gimlet eyes.

"Wow!" They chorused. "Buffalo Bill!"

William Frederick Cody sighed again, and wondered if killing Indians - even fictitious ones - was a lot easier than this was going to be.

'Buffalo Bill's Wild West' had come to Munich on a grand European tour, complete with Indians, squaws, rodeo riders, Annie Oakley, a mange-ridden buffalo and a hundred stampeding horses, it quickly become the talk of the coffee houses and horse troughs throughout the city. They had performed before thousands to foot stomping applause, stood in the presence of parliaments and barons; saluted the very bastion of the British empire her most royal majesty Queen Victoria and survived; re-enacted Custer's Last Stand so many times it became a byword for high adventure and pistol shots. Their stage shows made the fortunes on both sides of the Atlantic, most especially it made William Frederick Cody's fortune, and he wasn't about to let an unscrupulous hack novelist from Munich use his name without paying for it.

"But I'm telling you, Mister Bill, it is tribute to you," stammered Rabbi Silverstein as he stared the gunslinger with gimlet eyes from Iowa, "Its not meant to be derogatory – it's a eulogy!"

The Silversteins had hastily convened with half of the neighbourhood in the rabbi's parlour. The neighbours all stood agog, at the sight a real live cowboy who was - to their immense enjoyment - in the process of shooting more of the lampshades.

"Well, what kind of a catfish is this here Cossack you got written down," he paused as he thumbed his way through the pages, "where it says 'Seeing a golden opportunity the swarthy Cossack tore at sweet and innocent Esther's blouse..'"

Frau Silverstein fainted.

"Esther!" Rabbi Silverstein went to her side, "it's not what you think!"

"She's Esther?"

"Poetic license!"

"That's a lot of license."

"It's glandular."

"So I see."

"Now see here Mister Bill," protested the Rabbi, "it's one thing to complain about my dramatisation of your life and to shoot out all the lampshades in my parlour, but it's altogether unacceptable to ridicule my wife."

"Cody. Its William Frederick Cody."

"What is?"

"My name, it's not Buffalo Bill, that's just a nickname, my real name is William Frederick Cody. You've darn gone left that out from your book and all."

Silverstein appeared thoughtful for a moment. "So, legally," he pursed his lips, "legally, you're not Buffalo Bill."

"Eh, course I am."

"But it's just a nickname, you just said so."

Cody shot out another lampshade.

"Don't be trying any legal monkeyshines on me, or I'll be tying you to your missus Silverstein and riding the both of you out of town!"

"Oy! My god must really have it in for me today. That is a fate no man would wish upon another." The Rabbi pleaded with his hands. "But Mister Cody, Mister Cody, it's just a story, a little story, what harm could it do?"

"For your information," Thundered William Frederick Cody, "Warsaw is in Missouri, not this place Poland! I've been there, I know! This book is naught but fabrication!"

"Yes, yes mister Bill ..."

"Cody."

"... but it's just a story ... and nobody round here knows where Poland is either. And what kind of a name is Cody anyway? Buffalo, Buffalo - now that is a cowboy name!"

There was a vigorous shaking of beards from the neighbours.

"For the Torah says a man should be known by his good name." Rabbi Silverstein continued. "Unless of course, he's my uncle Vinny. Oy!"

The upshot of it all was the destruction of all the lampshades in the parlour and Rabbi Silverstein agreed to use William Frederick Cody real name instead of Buffalo Bill. In return William Frederick Cody agreed not to shoot Rabbi Silverstein or his wife, and that afternoon they were all given free tickets to the wildest roughest, biggest rodeo this side of the Catskills, where they were treated to the rare sight of the Emperor Wilhelm having a cigar shot out of his hand by none other than Annie Oakley.

Kepler's Moon

In which Tycho Brahe and Johannes Kepler solve the mystery of the Spheres.

As the sun began to set, the two men stood in the courtyard of the observatory. One old, near the end of his days, the other almost just sprung from the womb, both in awe at the majesty of the starry night. Before them, the garden of the observatory lay open like a map of heavens, each tree representing a star in its place; each path circled the topiaries like the orbit of planets. The two men walked slowly, discussing first stars, then directions and where to urinate. They were also both roaring drunk and had been drinking from mid-afternoon, now with the approach of night they were in full swing and could barely stand.

"I'm telling you, it's a star!"

"Looks more like a comet. I'd know, had one named after me."

"Does it have a tail? Is it moving? Can you see any witches dancing on it? Its a star."

"Witches don't dance on comets, I've never heard that one before."

They fell over for the ninth time, gave up and lay on their backs looking at the glittering eternity.

"Your nose is shiny."

This brought gales of laughter, as they both pointed at Tycho's artificial nose of gold and silver.

"My wife is pregnant again," he said glumly after another pause.

"Is it yours?" asked Kepler.

This was followed by an uncomfortable silence for a minute, and then followed by a roar of laughter.

"Mine or the dwarfs!"

The laughter continued for many minutes before their wives eventually found them rolling about in the leaves.

"We're drunk, go away."

"The emperor is waiting. He wants his astrological chart you promised."

"Give him yours, just cross the name out, anything. I'm too drunk for this now."

"I was hoping you would do it in person," the Emperor himself appeared behind the wives.

"Ah, your majesty, I was ..." Brahe stood up stood up too quickly and his nose fell off.

"Never mind," the Emperor sighed good-naturedly, "you can send one in the morning, once these vapors have cleared from your head. As for you Kepler, I'm surprised to find you in such a state. I had such good report of you ... ah wait, it was Tycho that gave me such report."

"And with good reason, your majesty," Brahe said, then fell back in the shrubbery before finishing his sentence, "my friend Kepler is just a wizard with numbers! You should see what he does with duodecimal numberings ... stuff."

"Tycho, I see your nose has fallen off again. Try stronger glue. Goodnight," and walked off discussing astrology with the wives of the two drunken astronomers.

The two remained lying on the grass for several minutes. The sky lit up with the night and the stars burned above them, beyond their reach and beyond their understanding, the light filled with the garden with a blue-grey light, which shimmered on the dew.

"That was the king."

"You think he is angry?"

"Nah."

"He seemed a bit annoyed."

"Nah."

"You're not worried?"

"I know a secret of his, and he knows I know. If you know what I mean."

"Nope."

Tycho looked at Kepler and pulled out a flask of white spirits.

"He's cheating on the queen."

"Everyone knows that."

"His wife doesn't."

"Ah."

They lay on the grass counting stars.

"If you're blackmailing him because of that gossip," prompted Kepler, "doesn't that mean everyone can blackmail him?"

"Sure, and anyone else who tries to blackmail him, just disappears or gets a dagger in the back." Tycho smirked. "I, on the other hand, do his horoscope and have predicted he will die the same month that I do. Now he is terrified I'll get run over by a carriage."

"Ah, double-blackmail, nice."

As they went back to contemplation of the heavens, a score of falling stars raced from one a point in the constellation of Scorpio and fell to Earth. Bursts of light exploded in their field of view, the fireballs streaked for a moment then trailed away.

"So he has no idea that astrology is complete twaddle?"

"None whatsoever. Kings aren't clever."

Later they stood up and dragged themselves back to the house. It was more of an observatory with a house attached. Science had yet to produce telescopes and astronomy was little more than an excuse to watch stars and have late night parties. Tycho Brahe was a genius at giving parties, and was famed throughout all northern Europe, from Helsinki to Amsterdam, and Oslo to Prague; nobody had parties as wild as Tycho.

"You know, you can hear some meteorites, as they fall," Tycho said as they staggered back.

"Can not."

"You can, on a very quiet night, I've seen the grass shaking and I can just hear something, which I assume is the noise from the falling star, buzzing in my ear."

"Angels?"

"Or witches."

The entered the house and made for the kitchen. The night just begun and they had many hours till the dawn rose, which left them with lots of time to do their work. This was the reason for Tycho's wild parties, he had to stay up late into the night to make observations, he was after all the royal astrologer and …

… and in the kitchen was a moose, and its head was in a vat of beer.

"What is that?" Kepler asked, never having seen a moose before.

"It's Rix, my pet moose, I brought him from Denmark."

"You have a pet moose? It's awfully large."

"And drunk. That's our beer."

The moose continued gulping the beer.

"Hey!" yelled Tycho.

The moose held its head up and stared at its master, then gave a threatening grunt.

"Ah, best back up. Rix in a bad mood."

"I..."

The moose grunted again, its bloodshot eyes glaring at them.

"...I think we had better run" said Kepler.

"No don't worry, Rix always does that..." then Tycho changed his mind and yelled, "RUN!"

The moose clattered towards them, its antlers knocking over

pans and scattering astronomers. They ran up the back stairs with the moose in close pursuit.

"He doesn't do this normally! He's a tame moose!"

"He's now a tame, drunk moose on the rampage!"

"Getting him drunk is how I tamed him!"

They panted at the top of the stairs while the moose snorted up at them from the bottom. Drool pooling from its mouth onto the cobbles, as it snorted and clattered about, crashing its antlers against the walls.

"There goes our supper."

"And the beer," said Tycho as he pulled out a flask of white spirits.

"What shall we do? Won't your wife come back?"

"She's the one having an affair with the king," he grinned then echoed, "Alright, what shall we do?"

"Work, I expect," said Kepler and looked at collection of astrolabes, armillary sphere and giant quadrants cluttered about the flat roof.

"Damn," muttered Tycho, "my nose fell off when we were running up the stairs. No time for work. We have to rescue the nose, before Rix steps on it."

They looked down at the moose, in the dark of the stairwell it was all but impossible to see. But Tycho swore that just beneath the vast bulk of the moose he could make out the glint of his artificial nose.

"Gods! I hope he doesn't defecate," Tycho said.

"What shall we do? We can't leave it there."

"You'll have to lure it away," Tycho pronounced emphatically.

"Me? I'm no good with luring, especially not moose!" Kepler protested.

"Two things. Firstly you're younger and faster. Secondly

you're my assistant. So start luring."

"So what do I lure it with?"

"It does like beer."

"Which is in the kitchen."

"You could try the wall."

The walked over and looked over the parapet, to the garden twenty feet below.

"Remember to bounce."

"I'm a mathematician!" Kepler protested again.

"Which is a far more dangerous occupation than most people would give credit." Tycho smiled, "Now, climb down that vine, go into the kitchen and grab a bucket of beer."

"But..."

"Do you know how much a good artificial nose costs? Especially if it's made out of gold and silver?"

"Do you mind if I ask, how you lost your real one?" Kepler played for time.

"I had a duel with Manderup Parsberg, using rapiers."

"The mathematician from Rostock? What about?"

"Who was the better mathematician."

"Couldn't you just sit down and do sums?" Kepler asked in amazement.

"Hey! I already knew he was the better mathematician, a duel was my only hope! It was Christmas, we were drunk, we were kids, we fought in the dark, and I lost." Tycho sighed. "He was also the better duellist. Now go rescue my nose!"

Kepler lowered himself over the wall and gingerly began to descend. He very quickly became sober as sweat covered his face and his hands became slippery. Yet, before he had gone halfway there came a grunt from the bottom of the wall.

"Hold up!" Tycho yelled down. "He's smarter than we

thought, he's waiting for you on the ground. Whatever you do - don't jump down! And wait there, while I go get my nose!"

Kepler stifled a sob as he felt his fingers beginning to slip. He had never thought he would spend his hours with the great master of astronomy by helping him recover his nose from a rampaging moose, a rampaging duck – maybe, but never a moose. He made a grab at another vine just as the one he was holding broke. His attempt only partially succeeded as the new vine stripped away from the wall and he fell through the air to come rest just inches above the moose. The huge creature gave a tremendous bellow then began to lick his feet.

"Go away! Shoo! Look tigers!" Kepler tried vainly to scare the moose away.

"There are no tigers in the North," Tycho called out from above. "It's alright, I've found my nose."

"It's trying to eat me!" Kepler yelled back, scrambling to find purchase on the wall. "Or my shoes at least!"

"Bad sign that, although I was expecting it later this month."

"What?!"

"Mating season."

A few moments later Kepler seemed to leap over the parapet of the observatory roof, and landed next to Tycho. Who was busy reglueing the tip of his nose back on, with a special ointment he always carried around with him, for just such occasions.

"That was fast. Where did you learn to climb like that?"

"Just then," panted Kepler.

"I brought some more beer," Tycho grinned, "we can start work now."

"And the moose?"

"It will sleep it off, or wake up some of the cows. We'll be safe up here."

The night sky was in full tumult as the arc of the heavens lit up with a billion stars. Their task was to plot the position of planets as they floated against this backdrop. Each pinprick of light in the firmament, representing a question mark about the nature of world and progress of humanity. Dressed in black cowls, trimmed with fur, their heavy cloaks insulated them from the cold, and if that wasn't enough a coal fired brazier glowed dimly besides them, as they blew on their fingers to keep them warm while writing down numbers on star charts and sipping cold beer. From time to time they would move the giant callipers they used to measure the arcs between the stars, and compared the differences between the hour glass, the hour-candle and a large clock that the Emperor had provided Tycho in his studies.

It was Tycho's task with his extraordinary eyes to pick out a star, or comet or planet as they slipped through the night, using the astrolabe for line of sight. In turn it was Kepler's duty to take note and a running calculation on the distances the stars were expected to move, then they compared the result against a giant armillary sphere that sat rusting by the door. Tycho was convinced his model of the solar system was superior to both the Copernican and the Ptolemaic systems, the trouble was the numbers always refused to add up. It was a slow, methodical progress and they kept their spirits up by telling tales and jokes.

"I make it almost a second of an arc."

"Martin Luther goes into a monastery…"

"Heard it," Kepler interrupted.

"How do you know what it is?"

"Luther says to the abbot, 'I'll have two nuns and a Hail Mary, to go.' and the abbot says 'I'm all out of nuns, but come back in two weeks because the Whore of Babylon is coming to town.'"

They fell over laughing.

"Okay, mark, and another second of arc." Tycho continued the measurements. "Okay, how about this one. The Pope goes to heaven, the gate opens up and there to his surprise is Martin Luther. The Pope says, "There's something wrong, surely your supposed to be in Hell.', Luther says, 'Is that so? Send me a letter, when you get there!', and slams the door in his face. The Pope bangs on the door and demands to be let in. Now Saint Peter opens the door and asks what he wants. 'I'm the Pope!' says the Pope, to which Saint Peter says, 'And so am I!' and slams the door in his face. Once last time, the Pope bangs on the door. The Pope's wife opens it and the Pope says, 'Whoops, sorry, wrong address.'"

At this point they heard a strange noise behind them on the roof and they looked around. There, its red eyes glinting in the candle light was, Rix, the sex starved and considerably drunken moose, staring very quietly at them.

"I think he's out of ale," said Tycho glumly.

"Maybe he couldn't find the cows."

"What do we do?"

"Whatever it is, do it very slowly."

"Wait, we're doing this wrong. I mean, you're the mathematician – what are the odds of a couple of astronomers being killed by a moose on top of an observatory in the middle of Prague?"

"That depends on whether you are the astronomer?"

"I think that's taken for granted."

"In that case," Kepler yelled, "Run!"

They bolted across the parapet and darted behind the giant armillary, but the moose remained at the door, breathing heavily through its nose, its breath appeared like dragon steam.

"Damn, that's a clever moose," said Tycho, "I was hoping we could run around it and go down the stairs."

"Will it eat us?"

"It's a moose."

"So, will it eat us?"

"Moose are vegetarians."

"So why are we hiding?"

"Moose are very angry vegetarians."

They waited for inspiration. They waited about an hour for inspiration.

"I think it's gone to sleep."

"On its feet?"

"Cows do, why can't a moose?"

"Its eyes are open."

"A sleep walking moose?"

Suddenly the beast gave out a bellow and charged across the roof to them.

"It's having a nightmare!"

They clambered up on top of the giant armillary, its score of planetary rings spinning in their orbits as they sort to climb to its pinnacle. The moose trotted around the base, bashing it with its horns trying to dislodge them, the stench of a rutting moose filled their air, as they held onto the top ring and pedalled with their feet to stay in place.

"Why the hell do you have a moose?" shrieked Kepler.

"It was for Christmas a few years ago. I thought, you know, for the kids."

"But a moose?"

"Well, it did grow." Tycho explained as they finally brought the spinning wheels to rest. "Rix was awfully cute and clumsy when we first got him, back then he only reached up to your chest, and the kids really loved him. Then he discovered alcohol and the household hasn't been the same since."

"And what do you do when he goes berserk like this?"

"Sometimes he sobers up, or falls asleep, or even just gets plain bored." Tycho pondered. "Mind you ..."

"What?"

"This is the first time he's been both drunk and in the mating season."

Kepler looked down at the spinning planets beneath them and had an idea.

"You know, from this perspective, it makes a lot more sense to put the Sun at the centre of the solar system, and put the Earth with the other planets."

Tycho looked down.

"Oh yeah."

At that point, the moose re-entered the conversation by attacking the armillary again.

"Where would we position a giant moose in the solar system?" Tycho burst out laughing. "With the comets or the stars?"

"Careful, or it may break the crystalline spheres," shot back Kepler laughing as well.

This only infuriated the moose, which redoubled its assault, in so doing; one of its horns caught on Jupiter and began to drag it around. The moose in its rage started running faster, which in turn set the giant armillary spinning wildly.

"Whoa! Whoa!" the two astronomers cried out as their world began revolving faster and faster.

Then with a tremendous crack, the moose broke the stand and set it rolling across the roof. The moose completely out of control spun its legs all a canter across the floor before finally falling down the stairs. Tycho and Kepler still grimly holding onto the giant model ball of planets, as it whirled about bouncing off walls and other astronomical paraphernalia, until it too fell to pieces and the astronomers flew over the walls and landed in the trees of the topiary below.

"Tycho," panted Kepler from his tree.

"Yes," a much bruised and battered Tycho replied.

"You sure know how to throw a party," at which, they both laughed and fell out of their trees.

Einstein and the Tour de France

In which a man, his bicycle and the Tour de France collide.

The Rhône, the Rhine and the Danube trisect Europe with a geometric nicety that Archimedes would have waxed lyrical about. For the Rhine, slow and majestic, works its industrial way North from the Alps, sealing the border between two ancient states whose only real difference was a taste for either Bratwurst or Crème Brule. The Danube, in turn, as it winds upon a long sinuous waterway to the east, is not so much a sedate river as a senile canal, passing through so many languages and different peoples as to be positively demented by the time it empties into the Black Sea.

Only the Rhône roars with vigour. Starting at an altitude of 1753 metres in the Rhône Glacier upon the Saint-Gotthard massif, it tumbles on its way to the Mediterranean in places it is a tempestuous fluvial wash sparkling with life. A gorgeous vibrant river, charging across France until it splits into the Grand and Petit Rhône and oozes through the Camargue delta, sliding into an expanse of marshy plain, filled with flamingo, feral horses and hordes of the most vicious mosquitoes in Europe. The Camargue once a wilderness, a dangerous place filled with outlaws and pirates, and in the very early days of the Tour de France, the Camargue was considered an impossibility, it was never included in the official route - until, of course, this fateful day.

The air was filled with the sweet scent of sunflowers and hay fever. Einstein rode along the bank of the river's, humming drinking songs and wondering about Love. This brought him no end of confusion, for women, wherever he met them, always seemed to have the same response to his overtures, a startled look followed by the clatter of shoes fleeing his presence. It was not his charm or lack of it, nor was he an unattractive man, indeed with his moustache and bicycle, he was considered one of the most attractive bachelors in Bern, until of course, women were confronted with his conversation. He seemed totally incapable of finding any subject of commonality, or topic of interest that was more interesting than cigars, beer, and bicycles. Not even the wildly fascinating subject of cuckoo clocks brought him any saving grace.

Indeed, he was considered the most boring man in Bern, and this was saying a lot.

The subject had kept him so occupied, as he sped along, that it only gradually dawned on him, that he was no longer in Switzerland, but had somehow passed beyond the border into France. He stopped and checked his visa. He had long given up on the possibility of getting a new German passport, and it was his Swiss visa was all that kept him from being deported to Russia, or worse still, Dublin.

About him, he noticed numerous people had lined the road waving flags and staring eagerly at him. He thought at once of South Sea cannibals, vanished missionaries in deepest Africa, and stranded polar expeditions that survived on nothing but huskies and Eskimos. So he hefted his bike pump and asked for directions.

"Bonjour, Monsieur! Le Tour!" An exuberant peasant smiled and kissed him on both cheeks.

"Ja, you don't say - is this the way to Zurich?"

"You are winning! Bon chance!" As two girls ran up and kissed him hungrily.

Straight away, his mind went back to thoughts of cannibals and their varied diets.

"I must be off!" He cycled furiously down the road, dodging from side to side, as groups of fans crowded the road, attempting to cover him with garlands.

"Allez! Allez!"

He looked over his shoulder to see if someone called Ali was chasing him. The crowds grew and chanted more vigorously. He started looking for cooking pots, and saw to his dismay chains of garlic being tossed in his path.

"Mein Gott! I'm the main dish!"

He charged on, looking neither left nor right and he suddenly entered the centre of a town and screeched to a halt. Surrounded by screaming fans, who rushed forward and carried him off to the main podium, throwing their arms about him, and kissing him with wild abandon.

"Is there a train out of town?!" he asked in fright.

"Bravo Monsieur!" A rotund mayor equipped with sash and wine bottle kissed him repeatedly.

"I like the girls you know!" Einstein defended his cheeks.

"Ha!" Laughed the mayor, "fear not, for today you shall be kissed by all the girls! Bravo, you have won a stage of the Tour de France! You are immortal!"

"Bravo! Bravo!" erupted back and forth across the town square, as a line of saint's statues was paraded from the church to be anointed with holy water and bicycle grease.

"Tour de France?" Einstein was very surprised. "Tour de France? There are guided tours for the whole of France?"

Instantly the crowd fell silent.

"You are not racing in the Tour, Monsieur?"

"Not so much not racing - as I'm late for work."

They deserted him, and shortly after crowded around a postman who also was the subject of mistaken identity.

By the time, the authentic Tour de France racers sprinted into the villa; the whole populace had completely lost the inspiration and gone home. The riders looked about in consternation, for they had expected a town of jubilant fans and were instead met by the ghostly silence of French town at midday. After a quick discussion they rode off in the wrong direction, having convinced themselves the entire Tour de France had missed a turn and got lost in the meandering conduits of the Camargue marshes.

Albert, after staring myopically upon the limestone walls and buildings, managed to find his way to the mayors office. The mayor had retired to his patio for lunch, yet after a moments consideration invited Albert to share lunch with him. The small matter of Gallic pride and the missing Tour de France was quickly forgotten over a glass of chardonnay and bouillabaisse de aiguilles.

"Monsieur. You are not French? Non."

It was not so much a question as an indication of a geological rift across the Rhine, this was no real divide for the mayor had spent some years as a waiter in Germany and spoke it well.

"Ach, no, I am German," Einstein replied, "sometimes, but not today, I have a job in the Patent Office, it's very good. I don't suppose you could tell me where it is, a large building with doors out the front that attack you?" He gestured with his hands to signify the enormity of the front doors.

"You are in the valley of the Camargue, Monsieur," a wry smile upon the mayor face. "Are you truly lost?"

"Lost? No, I mean if I go back the way, I came, then, … wait… yes, I'm lost…"

"No matter, we have a saying here in Camargue, -a man who loses his way in the Camargue marshes will drown.-"

"How is that a saying?"

"Well, it's true," the mayor replied with a shrug and a glass of Chardonnay, "and you can not argue with facts."

"What if the man - found his way back out again?"

"Never happened."

"What never ever?" Albert asked disbelieving.

"Not once."

"Wait, if someone was able to find their way out, then they wouldn't be lost, would they?"

"Ah, oui," The mayor contemplated this novel thinking. "I'd never thought about it that way."

"And how many people have you lost in the marshes."

The mayor pursed his lips and thought for a moment, "About five, I think."

"Do you know what happened?"

"Why yes, they moved to the next village."

This brought a very lengthy pause, as Albert considered the ramifications of this oddity.

"So, they ... didn't actually drown then?"

"Well, no." A shrug.

"So?"

"It's just a saying." With a smile and a mouthful of Chardonnay and langoustine à la crème. "I didn't say it was true. I just said it was a saying we have here in the Camargue."

Albert sighed. "Do you have any beer?"

"Alas no, Monsieur, you like beer, but this is the wine festival."

"Ach, ja, a great fondness for the bier, I have."

"Here then, have a glass of this, it is like, but without the bubbles."

Albert was amazed. "How do they take the bubbles out?"

A sly grin from the mayor, upon realizing the level of opponent.

"We have a machine for removing the bubbles, of course." A broad grin enveloped the mayor's moustache.

"Ach! The wonder!" Albert drowned in a glass of wine.

After lunch and several bottles of wine, the mayor

mischievously conducted Einstein on his erratic way, leading him to the edge of the town and pointing him in the direction of Tierra del Fuego.

"You must go that way, for about nine thousand kilometres. Remember to turn left at Brazil."

"Brazil?" Came a slurred and puzzled response.

"You do wish to see the world?"

This question was irrefutable, as indeed the young Einstein did wish to see the world, but at the back of his head the necessity of getting to work by eight-thirty was also a concern. As Einstein realised he was utterly lost, he stared about the medieval village.

Unbeknownst to him, it had not changed much for two thousand years, and in those millennia it had been conquered thirty-seven times by everyone from Hannibal of Carthage who once parked his elephants giving birth to the town, to Sancho the III, Bastard of Navarre, who held it for an hour before charging back to Navarre, even King Charles of Valois, had royally halted on the town square to use a chamber pot before sailing off to Sicily to be defeated by a bunch of sheep herders, and most bizarrely to a football team from Milan when a mafia fixed matched got out of hand and they actually won. Yet, in that time, the town had not once defended itself against anyone. It was a mystery as to how it had survived.

Meanwhile, the Tour of France had galloped off into the heart of the Camargue Marshes and were now fighting off some of the most aggressive mosquitoes known to biological science, so fiendishly belligerent the Geneva Convention had labelled them as weapon of biological warfare. In short order, all the cyclists had all dove off their bicycles into the relative safety of the marshy waters and covered their faces with mud to drive off the insidious insects. Soon to discover to their horror, the marshes were also habitat to the equally vicious Camargue leech, leading them in turn to leap back onto their bicycles and career back out of the Marshes.

It was this sight that confronted Einstein, as a group of screaming Frenchmen pedalling straight towards him, waving furiously about them and covered in mud and leeches. If he wasn't sure about getting to work on time before this, he now was completely convinced that this wasn't the main street of Bern, and promptly reversed his bicycle and vanished back up the road towards the Swiss Alps.

The pursuing riders, once they had shaken off the mosquitoes and blood suckers, immediately gave chase, - after all, anyone who could ride as fast as Einstein was at that minute - just had to be in the Tour de France. So began a chase that was to be remembered in the annals of the Tour, as the Grand Charge of 1903. This being the very early days of the Tour it was terribly disorganized, being French it was even more disorganized, and chasing an accident-prone patent, and navigationally inept, clerk from Switzerland only gave it a certain piquant. Einstein far out in front, found he was unable shake the hunt, and the cries from the spectators only served to encourage him not to be eaten, for he had heard about the eccentricities of the French cuisine, and it was not adverse to the occasional German hiker.

He still couldn't figure out who this 'Ali' was everyone was yelling for.

This Wild Hunt lasted through a burning day, and long into the night with spectators all the while blasting on hunting horns, Einstein in his vested suit and pants making breakaways with a horde of panting French determinedly in pursuit. It was at only dawn at the Swiss border checkpoint that any sense of sanity was finally allowed to prevail.

"Passport," came the inevitable laconic question from the guard, as Einstein flashed his visa as he wheezed past and continued on into Switzerland.

Behind him, a small drama unfolded as the guard screamed, "Halt!" at the Tour, thus bringing the entire Peloton to a screeching standstill. "Passports!!" No end of vivid haranguing of the guard brought any resolve, as they demanded to continue on their race, having seen the race leader disappear across the Swiss doorsill. The built up frustration was too much for the cyclists and they pushed aside the guard and forced their bicycles into the steps of the Alps, still seeking to catch up. Instantly, the guard blew his whistle and within a few moments the local militia, equipped with muskets and bicycles, joined the pursuit.

There were now three groups, Einstein lunging ahead having seen he was in the home stretch, followed by Tour de France who in turn had picked up the Swiss army. It was at this point in the history of modern Europe, that the idea of a truly international cycling competition was born. This race continued all the way up to the front door of the Patent Office in Bern, where Einstein in a marvel of geometric manipulation was able to pass straight through the revolving doors of the office and disappear via the elevator to upstairs, as the entire group of pursuers collided into a painful heap and the race was declared over as they were all arrested for trafficking bicycles across the Swiss border without a permit.

This small moment in history also insured the neutrality of Switzerland in both the First and Second world wars, for the Swiss having seen the French could invade any country armed with only bike pumps and Peugeot bike frames, decided they were no match for this military goliath and remained forever indifferent to the possibility of war.

Newton and the Case of the Dancing Fools

Newton was once a copper.

It was a warm and blustery night, not the sort of night great crimes are committed, more of the type of night where everyone goes to the pub for a friendly beer and a game of Skittles or Dwyle Flonking. This however would not deter the greatest detective of the Enlightenment, Sir Isaac Newton, who positively detested Dwyle Flunking. Partly because he was always on the trail of arch-villains; or perhaps because it was beneath his Olympian mind, but mostly because he was never invited to play.

Dwyle Flonking is a game that requires some explanation, indeed it requires considerable explanation, and in fact it needs the type of explanation that leaves editors of the Encyclopaedia Britannica in sobbing tears. Suffice to say, it entails a group of grown men, standing in a 'girter' or circle, around a man in the middle who is the 'flonker', who stands still with a broom handle. The broom handle has a rag soaked in stale beer tied to one end, and with this the 'flonker' attempts to hit the 'girter' as it dances around him. There see, I told you it needed an impossible amount of explanation, and I kid you not – it is a real game.

Newton's piercing eyes took in every detail of the well-lit alleyway. He frowned. He preferred dark alleys with lowlife scum and crimes of passion. His brilliant mind realised this was another fruitless search for crime, but then again a schoolboy dangling a yo-yo could have told him the obvious. He was years ahead of his time, so far in fact; he had seriously considered inventing Time travel to get out of this god-forsaken age. Newton was a criminologist, a sleuth, constable for the crown - his friends called him a loon and couldn't understand why he didn't stick with natural philosophy.

Dressed in a dark cape that only made him stand out more in the bright full moon, he prowled along the edge of the street, his nerves stretched near breaking point. With pale white skin and walking stick, he seemed far more akin to a Grim Reaper than a detective on the hunt and this had lead him to be questioned on more than one occasion by a genuine bailiff of the village of Burton-Coggles.

He suddenly got down on hands and knees to examine a pile of earth that turned out to ox manure, to which he sagely deduced it was manure. He smelt it, probed it with a stick and examined it with a lens. It remained cow manure.

"Damn your eyes!"

"E're now," the bailiff in his night patrol appeared in front of him carrying a lantern, "oh it's you again, Master Newton," and wandered on.

Newton remained perched on the ground like an enormous toad staring at the cow manure, convinced it held the clue to some crime. The crime may not have been committed yet, but if he could just piece together the puzzle, then he could catch someone – anyone in the act.

There was a very good reason for Newton's eccentricity, although 'eccentric' may not encompass the full range of his antics, perhaps 'daft' or 'mad' is more descriptive. Anyway, the reason lay in that the English crown has a habit of rewarding great men with nominally useless positions in government, this ensures they would be kept out of harm's way and also makes certain they can do no harm. A typical position was, and still is, Lord Warden of the Cinque Ports, it has no political power and entails wandering around the ports picking up taxes and making sure the castles aren't stolen. Equally, the steward of the Royal Aviary is a position that carries no greater duty than feeding the royal pigeons. There are very sound political motives for doing so, for a man who excels in the public eye is often a grave threat to the stability of the crown and can occasionally rise to the position of a demagogue without careful handling.

In Newton's case he had been awarded, for his tremendous and extraordinary gifts to science, the position of Warden of the Royal Mint. Unfortunately, Newton had convinced himself that he really was supposed to police and guard the safety of the crown and it's coinage.

He had become a Copper.

Or at least he thought so, everyone else was astounded when he had begun to arrest people and put them on trial.

"Isaac!" The chancellor of Cambridge had recently upbraided him as he met him in his private chambers, "You have to be kidding, you're not meant to go out and deal with … with … commoners!"

"Tis my sworn duty to uphold the law, and by that sacred oath that I have taken, thus shall I do."

"But they're … why they're criminals. You can't mix with criminals, that's what bailiffs do. It's unheard of!"

"So was a Theory of Gravity before I came along."

"My dear friend, you must see there is a whole world of difference between a treatise of mechanics and arresting people for being drunk and disorderly."

"That I do not do and you are misinformed. For those whom I have apprehended have committed heinous acts against the crown, including larceny, counterfeiting, murder and possibly Dwyle Flonking."

"Wait, you caught a murderer?"

"Well no, in fact he was acquitted of the charge, but he was acquitted because of a technicality." He snarled. "Damn the eyes of those lawyers."

"What was the technicality?"

"He wasn't there."

"Doesn't that mean he was innocent?"

"Technically, yes, but that was the only reason."

"Who did he murder?"

"A cow."

"A cow?!"

"He was a butcher. Hence he had motive and means. All he needed was opportunity."

"But a cow?"

"A murder is still a murder, doth it not say in the Bible that to slay a sparrow is to sin against the laws of God?"

"Actually, no, you made that one up."

"I did? Oh, well it should be!"

"Oh poor Isaac, I fear your mind is cracked from the strain of all these years of reasoning. I beg of you stop this nonsense, before you yourself become a tragedy."

Newton glowered at him.

"This I can not do," he said, "for I have sworn a solemn oath to his majesty and this I will seek to uphold."

"Yes, but nobody expects you to!" The chancellor cried in exasperation. "Not his majesty, I certainly don't, nor for that matter the criminals themselves! The Warden of the Royal Mint is just a title for pity's sake! A piece of paper so we can reward you for all your scientific works. You're actually not supposed to do anything!"

"I disagree," Newton said gravely.

Hence, Newton had become a sleuth hound centuries before it had become fashionable. He did however go about in true scientific fashion, collecting evidence, interviewing suspects and arresting hapless bystanders.

No magistrate of the crown was able to dismiss the evidence, as the whole idea of scientific evidence was so new, that no one quite knew what to do with it. As a result Newton was developing a near perfect record for arrests and trials, however in case after case it was found the suspect was simply nowhere near the place where the crime had been committed. He was also developing a reputation amongst the magistrates for needless prosecution.

"Your honour," he once said in a trial, "if it pleases the court, I wish to bring forth a shoe as a witness."

"No, Master Newton it doth not please the court one bit. For the court does not understand how a shoe can explain the nature of this crime."

"Your honour, the shoe is to be a mute witness and will address the crime by virtue of it's being present at the crime."

"Master Newton, a shoe is not a living being, it hath not a soul and can not speak. So, it can not be a witness."

"Your honour," Newton said, holding up his shoe witness, pointing to its hobnail bottom and then to its tongue, "it hath a sole here, and a tongue here. Thus it is a pertinent witness."

To this the judge had no reply, and a hobnailed boot was sworn in and gave legal testimony against the shoemaker who was repairing it.

At the present moment, Newton was searching for a crime in the light of a full blue moon. He poked the cow manure once more, just on the off chance a clue might spring forth as Adam from the earth, but alas it remained uninteresting dung, as no doubt Adam occasionally wished he too had remained. Newton rose and walked along the dark passages of the streets hoping for something unusual to catch his attention and had been doing so since the sun had disappeared into the hills. He believed in preventative policing, not only catching the criminal in the act of their crimes but before they had even thought of it. It required a serious amount of concentration to discover crimes which had not yet been committed, a quantity of which Newton had in such abundance, he was starting to wonder if he was prescient.

Indeed, his level of extrospection was so high he could have solved the case of Christ's remains missing from the tomb after the crucifixion, except of course; it was already a case of legally sanctioned body snatching.

The pale shadows of the night shrank and fell away as the moon rose to its full height above the tiny village of Burton-Coggles. The rows of houses with their thatched roofs and half-timbered houses, wattle and daubed plaster stood like truculent housewives nattering away. It was a quiet place, one of those unknown villages that have never seen the stamp of war, rarely the gaunt stare of famine and hardly touched at all by the plagues that had ravaged the rest of Europe. Its only real claim to fame was Newton's distressful excursions into the law enforcement. For the most part the folk of Burton-Coggles wanted nothing more than to farm, drink and breed.

Newton paused again and glared at the happy voices coming from The Hare & Hounds Inn. He knew that anyone of them could perpetrate some hideous deed tonight, although they might need a little encouragement. Irritated he turned away and followed Chestnut Lane out to the remnants of ancient woods beside Burton-Coggles, there he stood down and stared at the sleepy village wondering why he had ever bothered to patrol. Then in the distance he heard a yelp of pain as someone cried out for help.

"Aha! I knew it! Tonight I shall prove them wrong!"

He knew this wasn't true, but he also knew he had proved them wrong on so many occasions, that this had to be another and was about to run into the darkling woods to carry out the process of the law, when he realized he was going to need back up. This was in an age where the nearest thing to back-up was a speeding wagon. He turned back into the parish of Burton-Coggles and grabbed the bailiff.

"Ah, er'e now master Newton, there be no need for grabbing a fellow like tha' in this here night."

"There is no time for this, a crime is being committed, and I charge you with the law to follow my instructions and apprehend a culprit who is 'In flagrante delicto'.

"Ah, now, he be 'en fragrant delicous' - tha' sounds more like a desert than thar criminal."

"This is no time for idle chatter."

"Ah, middle of the night be the best time for idle chatter, I would h'ave thought."

"Look man! A crime is being committed; you are the bailiff of this Burton-Coggles! I am the Warden of the Mint. I instruct you to follow me!"

"Ah, well put it like tha't, I'll just go get me good lantern, can't go out in a night like this with a lantern, tha' nae be me best, can I not like."

"Damn your eyes! It's a full moon! You don't need a lantern."

"Ah. Right."

They set off banging on doors hoping to raise other citizen, but this only brought the occasional opening of shuttered windows with cries of: "E're bugger off!"

"What is wrong with these people?"

"Ah well, we never had any crime before."

"What never? Then why do they need you, the bailiff?"

"Ah, hmm, I don't rightly know. Every village has a bailiff; they also have a fool and a witch. I mean what does a bailiff do?"

"Great Zeus in Heaven! Don't you know?"

"Ah … no."

"Well you arrest people, of course! And after they have been arrested, they are confined in a gaol until such time as they may brought before a magistrate and a trial be conducted."

"Ah, well your lordship, that's all well and fine, but we ain't never had any crime afore. So I ain't had no practise and all."

"Aha! I knew there was something wrong with this place. Has no one has ever been arrested."

"Ah well, thar was Tommy Cowper."

"Was he arrested?"

"No, but some of his jokes were just criminal."

"Give me an example."

"Two elephants fall off a cliff.. boom.. boom."

"I don't understand."

"Ah, I thought you would say that."

"We have no time for this, to the woods!"

"Ah, to the woods!"

They ran on, the bailiff stumbling along as Newton flowed from like a bat with his cape rippling in his stride.

"It was near here." He whispered. "Hark now and listen."

A bird called softly from the dim trees.

"Hear that?"

"Ah, that be a witch."

"Don't be a fool man, there are no witches. That is but a pastoral fallacy left over the Dark Ages. This is the Enlightenment, we have cures for disease that no longer need witches and black cats, modern medical philosophy uses leeches and bloodletting – we even have diagrams!"

"Well, we burned a witch last week. I wish someone had told me before."

Newton turned on his attendant and grabbed him by the

lapel.

"Did you say you burned a witch last week?"

"Ah, aye, harvest festival happens every…why every harvest I expect. We burn's a witch. Been doin' that' for longer than I know of."

"Great Orbit of Jupiter! That's murder!"

"Ah, is it? Funny I thought it were just a bonfire."

"Bonfire?" Newton almost shrieked in amazement.

"Ah well sir, it's not like we killed anyone."

"Then, you scoundrel, who did you put up on the bonfire?"

"Ah, we makes the witch out of wood, like we always do, tie together swathes of twigs in the shape of a man and puts it up top of…"

"Never mind," Newton sighed, "so what makes you think that noise we just heard was a witch?"

"Ah well, birds grow from trees, like apples, so stands to reason they be made from twigs as well…"

"Silence, you blithering oaf."

"Ah yes, your lordship," the bailiff coughed. "Shall I arrest the tree now?"

"Oh, for the love of justice. Tell me, you son of an empty headed popinjay - are you the only bailiff in this accursed community?"

"Ah, lets see, umm."

"Well?"

"Ah, yes."

"Then I charge you in the name of his majesty to just shut up!"

"Ah, have you met the majesty?"

"Yes, that's who knighted me! Now shut up!!"

"Ah."

They continued into the thicket of trees, the leaves dappled with soft moonlight, and the undergrowth seemed to scurry beneath their feet. There was another yelp of pain followed by a chorus of laughter.

"By the Crescent of Mars!" whispered Newton. "They are torturing someone!"

"Why be they laughing?"

"Fiends!"

"You mean like witches?"

"Worse! Anabaptists! Quickly, before we are too late for the poor soul. Have you a sword?"

"Ah, I've got a lantern."

"Use it wisely."

They charged into murk of the Lincolnshire Wold, screaming out the king's name to cease and desist. What they hadn't expected to find, was a grotto of trees with a group of men holding torches and dancing around a blindfolded man, who held a stick with a rag soaked in stale ale, who was trying to strike the dancing fools.

"Dwyle Flonkers!" cried out Newton, said like that it was seemed almost an insult.

"Oh, e're now, tis mister Bailiff and friend," said the mayor of Burton-Coggles. "Ye come for a game of Dwyle Flonking?"

"It's the middle of the night," said Newton, "and you are all out Dwyle Flonking? Is this whole village stark raving mad?"

"Oh, e're now," said the mayor, "there no need for that master Newton. We just be like hav'in a bit of fun, we said to ourselves down at the Inn, lets go dance round the great circle in the forest and have a bit of Dwyle Flonking. Just like in them old days."

"Have you not wives and family to go home to?"

"Oh, tha' be one of the reasons we came Dwyle Flonking."

"Why can't you play cricket like normal people?"

"Can't play cricket in the moonlight, master Newton, we'd keep losing the ball."

"But.. but.. but... why play anything in the middle of the night on a full moon?"

"Oh well, we don't want to look like daft idiots - I suppose, night time best for doing daft things."

A chorus of ayes and oohs mumble about the circle.

"Oh, so then master Newton, will you be joining us?"

"Of all the insane, mad, downright bizarre folk I have every had the misfortune to stumble across, this village is without a shadow of a doubt the most backwards and deluded group of idiots that has every existed. I cannot express the shame and outright sense of indignation I feel, that a group of Englishmen would conduct themselves in such a primitive if not animal fashion. You sirs, disgust me!"

"So, is that a yes?"

It was too much for Newton and he crashed off through the underbrush swearing something about the Holy Trinity and the redemption of lost souls.

"Thought he'd never go," said the mayor with relief. "Right then, get out the coin stamp, we got to have this counterfeiting done by sunrise, before this damned new coinage of master Newton hits the banks."

Einstein in Dublin

Top of the morning to you,
they don't really say that – A. Einstein's memoirs

"Top of the morning to you, Michael."

"And the rest of the day to yourself. Buck Mulligan."

"Would yee be happening to have a bit of that' thar' India Tea."

"Why yes, Buck, indeed I do, I have a little bit stored away in that nine hundred pound barrel, which your resting ya' arm on ya' blind idjit."

Einstein walked into the shop, lost as ever, looking for directions. He had come to Dublin on his way to America, after mistakenly boarding a fishing boat in Bremen and falling into a hold, where he was lost for five days until it docked in Dublin, to the great surprise of the crew and the relief of Einstein who had subsisted on raw herrings for the duration.

"Top of the morning to you, sir," opinioned the two Irishmen.

"What is?" Einstein looked up, and stared about the ceiling of the dockyard store.

"The morning sir, top of the morning to you."

"What's at the top of the morning? Lunch? Do you serve lunch, I'm famished, oh I do hope you're not serving herring."

"Herring? No sir, this is more of a dry goods store."

"Oh that's good to hear. Do you have sandwiches?"

"You'll be wanting the pub up the hill, they have a foine counter lunch, they do."

"I can take ya' the way there sir," said Buck Mulligan, a large man with a brutally mashed face, who offered his paw to Einstein, it seemed to swallow up Albert's whole hand and crush it with a unrestrained friendliness, which brought a moment of excruciating pain Einstein who thought he was being mugged.

"Oh now," intervened the shop owner and pushed Mulligan aside, "take no notice of him, he's a big lad, but he's not the sort to kill you."

"Oh, no," rejoined Mulligan and grinned, "I've never met a man I've had to kill. I'm not the killing sort, at all. Oh no, not like that Stephen Dedalus, he'd break ya' neck as sure as look at you. He's a terrible one for the blood sports. He is."

"That's good to hear," Einstein wrung his fingers.

"Right then, lads,' now then, Buck, ya' was askin fur' a pound of that thar' India Tea, you was."

"Do you have any coffee?"

"Ya just asked fur Tea ya daft idjut, what is it Tea or Coffee? I ain't be givin' ya all day."

"Well," Mulligan rubbed his two day stubble, do ye' have a dram of the mother's milk?"

"I dinna have time for this, I work a three hour week and I'm worn to the bone, git out ya damned fiend frum hell! I'll not be takin' no more of ya' nonsense frum yee." The store keeper Michael then grabbed both of them by the elbows and hurried them to the door. "And don't come back! Ye son of a blind whore from Brussels." Slamming the door behind them.

"He's a lovely man, I couldn't have asked for a finer father," grinned Mulligan, "Now then, me foine lad, lets get ye' to the pub and ye can buy me a lunch, I don't moind. I'll be happy fur ye' to pay. Ya' a lovely man, foreigner, I take it. Don't get many of them round here, not like and all we have much calling for' tha' sort of thing and all. You're no English, cause if you'r were, ya' lovely fellah, I'd be crackin ya' skull faster than Dedalus biting the head off a hedgehog."

"No! Not English, German in fact," Albert broke through the

Celtic divide, "well, was, there's a little problem with getting my passport back."

"Well, that's lovely to hear, and here we are ..." He waltzed Einstein through the door and into the bar, "I'll be havin' a porter, y'a a charming man, best we be asking Molly about the counter lunches, she's a grand lass, but best not be givin' her the eye if Dedalus is around he' ... ah, the'rs the lovely man himself, Stephen, over here, I met a new fellah, he said he'd be buying drinks all round, ar' heres to you Molly, how ya' sister, I h'eard she was goin out with O'Connell... damn sham tha' ..." Mulligan swam across the tavern, seemingly immune to the insults that were hurled at him from about the bar, to make his way over to the darts board, leaving the intrepid Einstein staring up at a man who was even uglier and more facially damaged than his wayward guardian angel had been. There was a bruise above his left eye and a fresh cut.

Dedalus the Irishman rose about two feet above Albert, his face was seemingly crafted out of driftwood, a dozen scars laced around a nose so cauliflowered by pub fights and street brawling it had shifted completely to the left by an inch, two beady eyes which glowed with undiluted hate stared down at Albert who found himself holding a ham and cheese sandwich and a 'foine' pint of blackest porter care of the lovely Molly.

A sense of self preservation that occasionally expresses itself in human race in moments of extreme danger allowed Einstein to offer his ale to Dedalus. The towering Irishman took the ale in his left hand, and now found to his immense pleasure that both hands were filled with ale.

"Oooh, you' a lovely man, top of the morning to ya." He smirked and sipped first one beer and then the other. "Ahhh.. a good craic, that is, if I had been getting up this morning an' I had stop to ask meself if a lovely man would be giving me a free pint I never would have believe meself, and I'm a right clever man you know. Who are you then? You look foreigner, you'r not an English dog are you, cause no offense to you and all but I be twisting your head around backwards and pulling it out ya' arse if you were even the slight bit of an Englishman."

"German," Albert coughed. "Germany."

"Are ye'now," he pursed his lips thoughtfully, "well, now I don't believe I've ever heard of that part of West country afore, must be near Mayo, ah well I be learnin' something new every day." Then grinned and wandered off, as Einstein left go a slow breath and crumpled against the bar, then hove into his sandwich.

"Oh don't be mind'in that tha' Dedalus, give him a walloped on the noggin an' he's just a sweet mutton an' all." The bar owner, an amply buttressed Molly Bloom advised him, as she managed to heft a club in one hand and carefully polished a glass with the other. One could still see a spot of blood from Dedalus's head on the club. "Tha' will be two pounds four shillings for the sandwiches and the porter you just bought the bar."

Albert offered up a handful of pfennigs in recompense, not willing to argue with a woman who was romantically linked with anyone called Paddy or Dedalus.

"Ah, what would these be now?" Molly squinted at a pfennig, "Must be that new currency their' been talking about," and tossed them in the till, then handed him five pounds change. "Now, then you'd be new in town, I'm guessing, and will be needing a place to lay your head, wouldn't you, well I've rooms here.

"Is this America?"

"Amerikay, you ask?" She smiled enthusiastically, "No, you'd be wanting the next country across the Atlantic ocean, ther' other side of County Clare, we tend to call this place Ireland."

"I thought this would be America, I am emigrating you

know."

"Is that so? Who would have thought a lovely man like you would be goin' to Amerikay, you must have the ambition I take it. A foine thing it is to be havin' the ambition and all. Aye, but ti's not for the like of Molly, I'd be tellin' you, and I just did."

Her immense frame floated down the bar, swabbing down spilled porter and possibly blood from Dedalus's head.

Another voice appeared at Einstein shoulder and he looked around askance.

"Oooh, hello there me lovely fella' I'll be thanking ye' for the porter an' all, and here's me not even known' ye name and all."

Albert looked in some amazement, to behold a broken man with dark tinted glasses, sad moustache and sucking on a pipe. "You're not foreigner are you, for you see I've been looking for a foreigner to fit into my next book, and I have the feeling that you might be that falla', Joyce is me, me friends call me James."

Einstein could only nod his head in his usual bewildered response.

"That's lovely to hear. Now can y'ou tell me a bit ab'at y'ouself, tis fur me next book and all. For y'ou see this is a small town, Dublin it is, and I don't get much of a chance to be meeting any of y'ou foreigners much besides those who come off the boats day tripping from Munster and even if I was to be wantin' to hav'in a mention of them in me next book, I'd not be needin' to, as all the world knows that if y'ou are wantin' something done, y'ou had best not be askin' someone who sailed down on a boat from Munster expectin' to be givin' a job as if it was they're god given right. Not that I be havin' a problem what with all these talk of seperation from th'e damned English and they' wish to impose upon us what they would be thinkin' is their god damned right of god save the King, may he sink in the English channel next time he goes fo'r a swim, which is isn't likely to do and all, what with his syphilis and all. Now then what was y'ou name and all?"

Einstein offered him the remains of his sandwich.

"Ah, now, y'ou a lovely man. I'll be sure to be putting y'ou in me next book and all." James Joyce wandered off munching the sandwich scribbling a note on his cuff and he went.

Einstein back himself against a wall and stared about in fearful confusion. Was this whole town mad? He asked himself, as yet another group of men erupted into singing around the fiddlers and penny whistle blowers. He wondered if it was simply bad luck that everyone he had met was either talking complete nonsense or threatening to crack his head open, after all, a whole town couldn't be mad, could they?

He went once more to the Molly and somehow managed to buy another round of drinks for the entire bar. The conversion rate of pfennigs to English pounds hadn't changed, indeed it seem to go up, leaving him with an even fuller wallet than when he walked in. She assured once him that Dedalus was as gentle as the lamb of god. This lamb of god hurriedly stomped across the room to remind him that if he even blinked in the general direction of Molly he would find new ways to remove teeth that hadn't been invented by dental science, at which Molly removed a tooth from Dedalus and told him to leave the paying customers alone.

Eventually Einstein was able to find a sheltered corner of the tavern, and hunkered down over his second round of beef and tallow sandwiches - but not for long.

"Hello, again, me foine fellah," prompted Joyce as he slid onto the bench opposite Einstein. "I see y'ou eatin' some of the foines't tallow this side of Wicklow, now if tha't is not a reason to be celebrating, then I wouldn't be know if it was Wednesday or Sunday Lent before a summer filled with downy breezes and long ale in the shade of a half forgotten relic of a roman burlesque."

It was at this point, Einstein realised Joyce was possibly suffering an inability to shut up, and again handed him the remains of his second sandwich.

"No now, you're being too kind you are," Joyce smiled and pushed the sandwich back, "I've nothing to offer in return but me wit and me genius and who would you be, what with your moustache and your foreigner tongue."

"Perhaps you've heard of me," Einstein coughed, "I'm the inventor of the theory of Relativity."

"Oh that fella would be yourself," Joyce picked up with interest, "no never heard of you. Where are you heading, across the sea I would wager, if I'm not knowing what I was thinking I wouldn't be sayin' it in the first place, after-all never a dance in the day without a romance in the morning, far across the ancient seas in search of adventure."

"Er," Einstein's brow wrinkled in exasperation at deciphering the rational from the babble, "yes maybe, I suppose, you mean like Ulysses and the Iliad."

"What was that?!" Joyce spectacles shone like bottle bottom glasses on a ferret trapped in a jam jar. "Like those ancient voyagers the Greeks did travel, when away to the Troy did sail Agamemnon and the warrior Achilles. What an idea! Of course! A modern interpretation of the Iliad, but set here in Dublin! Man you're a genius!"

"Um, not quite."

There was a crash an Buck Mulligan fell on the table and grabbed the ale, before Joyce ran out the door and Molly Bloom screamed, "Mulligan put down the paying customers!" Inadvertently saving Einstein from being strangled, who also ran out the door and caught then next boat to Amerikay!

Einstein and his Violin

*In which Einstein's love of the violin is born,
and music is put back a thousand years*

Frau Einstein dragged the wailing young Albert along the Munich Strasse, his arms and legs whirling about, grabbing onto any object or pedestrian that came within reach.

"But Liebchen, its only … ach!"

The young Einstein grabbed the beard of a passing Rabbi.

"Ai-yee!"

"I sorry Rabbi Bernstein …" exclaimed missus Einstein, "it's for a music lesson."

"A music lesson? A music lesson?! Is it for the Barber of Seville?!" muttered the Rabbi through his tattered beard as he scurried away.

"Now then, stop this screaming, Liebchen, it's only to see a music teacher."

The Einstein child only screamed louder and wrapped himself around a linden tree screaming to the high skies, this brought a police officer huffing up the road, who stood before them hands on hips.

"Mein Gott! Is your child kaput?" Came the statutory question. "For, Madam, it is forbidden for children to clutch trees in the City of Munich."

"No! No, it is the music lesson," placated Frau Einstein, "he has never been before."

"Ach, the terrible music lesson," smiled the kindly officer as he stroked his moustache, "hmm, let us see if we can get you there," and leant down to Albert's head to blow his State regulation police whistle into his Albert's ear. Albert fell back with surprise and would have fled down the Strasse, if the sympathetic police officer had not benevolently collared him and hoisted him clear off the ground. "Now, then. Where is your music teacher, hmm?"

They dragged the squalling child up an ever meandering stairway, as hygienic white walls surrounded them, the despairing screams of Albert bounced up and down the stairwell. Gustav Mahler pushed his glass up his nose and stared down at the noise machine, which had erupted into his studio.

"The child who will not play... hmm?"

The five-year-old Albert gibbered back up in fear.

"But he is most musical, Herr Mahler," began Frau Einstein, "he has a triangle."

"A Triangle no less," leaning down, "and can you divide music into angles – Hmm?"

This brought a puzzled stare upon both Albert and his mother, who looked at each other askance.

The Toddler decided he had had enough and made a break for the door, only to be scooped up by the waiting police officer. "Ach, nein, little one, now is the music lesson, and ... oof!" As one of Albert feet discovered the law of groin injury. "Ach! A fighter is this one..." pulling out handcuffs to arrest the perpetrator.

"Ah..." interrupted Gustav, "no need for that, with my teaching methods the little one will soon have a musical ear."

Mahler then dodged an accurately thrown triangle that smashed through a window behind him.

"Hmm ... Perhaps just this once. Put the bracelet on his ankle and lock him to that chair. Hmm ... and that will be two pfennigs for the window.. Frau ... um?"

"Frau Einstein, Herr Doctor Mahler," beamed the jubilant mother, "and this is Albert. My pride and joy."

An eye-goggling scream stopped their conversation, like a blind drunk bridesmaid landing head first on the wedding cake, this was followed by a mild chuckle as Gustav and the police officer shuddered. "Well, that was no Ode to Joy ... hmm ... was it?" Now then young master, let us see what you can do with this ...shall we."

Handing young Albert a trumpet, this promptly flew out of the young master's hand, bounced off the police officer's head and went through another window.

"Ow!"

"Not into brass, I take it."

Young Albert bit Mahler on the knee.

"That will be two more pfennigs, Frau Einstein. Well then," Mahler said hefting an oboe, "try this ... now the oboe has a noble history ... oh ... hmm not so noble it can not be used for cleaning windows. Well, that will be two marks, Frau Einstein, a good oboe is hard to find, and another two pfennigs for that window. Hmm...I do seem to be running out of windows."

"He is most musical Herr Doctor, why only yesterday I heard him screaming."

"Screaming? As in - screaming a tune?"

"Oh well not quite that far advanced, Herr Doctor, but one day!" Frau Einstein beamed.

"Hmm." The good director of Munich Opera Company nodded agreeably. "Screaming you say," and picked up a violin. This time instead of simply handing it to the young tornado, he lovingly screeched the bow across its strings and wandered over to the window, all the while loosing a cacophony of screeches.

Young Albert froze in astonishment at this tremendous caterwauling, and then strove to follow Gustav Mahler, dragging the chair shackled to his leg behind him, his hands leaning out to grab the wonderful wooden box.

Mahler turned as he screeched and seeing the boys enthusiasm smiled indulgently, "Hmm?" proceeding to walk back across the room circumnavigating the boy, he released a sound not unlike a rucksack of wild toms skirmishing over a plump dormouse.

The police officer who felt his teeth melting in response to the screaming banshee, "Ah. Ooh, I must run along," and sidled out the door.

"So, little one," holding the violin just out of the boy's reach. "Does this interest you?"

"Waaaguh!" young left off a scream that would have done justice to a dying Paleolithic wildebeest.

"Hmm, a little off-key, but he has some potential." Leaning over indulgently Mahler handed the boy both the violin and its bow.

Once in possession, of what would one day be condemned by the League of Nations as a psychological weapon, the young Einstein stared mystified at this musical eviscerator, then he look up at Mahler and poked him in the eye with bow.

"Aeeeei! Mein Eye!" screamed the master musician dancing around the room, "Wagner in Himmel! It hurts! Ach no ... try playing a string! The poking of eyes is verboten! That will be twelve marks, Frau Einstein, I know a good doctor."

"Yes, of course, Herr Mahler," radiated the maternally blinded Frau.

Then, standing back and holding a cello for a shield, the maestro instructed his protégé in the finer points of his art.

"No, no ... don't bite the varnish!"

That afternoon, refusing to acknowledge defeat in the face of certain failure, the good Mahler had taken his young Wunderkind to the Munich Symphony Orchestra, which was practicing for the coming Wagner festival. Frau Einstein walked behind them carrying the chair that was still fettered to the boy's ankle.

"This is the orchestra pit," began Mahler hoping to enthuse the boy, as the entire orchestra stood up and bowed to their conductor.

"Waaaugh!" screamed the boy.

"A Flat," attempted the lead violinist.

"Nonsense, that was definitely an E Sharp," riposted second violinist.

"I would have called it A," intoned the bassoonist. "Neither Sharp or Flat."

"Rhubarb!" "Rubbish!" and the occasional manuscript on musical theory flew about the pit.

"I'm telling you it was an A Flat," insisted the lead violinist and scrawled a note on his violin.

"Waaaugh?" interrupted the boy's larynx.

"There see ... definitely an E Sharp!"

"Was not!"

"Was too!"

"Not!"

"Was too ... too!!"

A crescendo of insults and musical terms exploded across the symphony pit, flurries of sheet paper cascading from music stands as the violin section started fencing with flats and sharps.

"Gentlemen! Gentlewomen! Gentlepeople!" Called their conductor Mahler. "A little diminuendo... if you please! dolcissimo, pianissimo." Then in a dramatic voice. "Silence!.. or ..you .. will .. all .. be .. playing .. in .. Bratislava ... again! And you know what that means!! ... Better - much better .. today we have a special guest."

With a flourish, he brought forth the boy, the mother and the manacled chair.

"A chair?"

"Are we getting new seats?"

"Is it oak?"

"No!" the bespectacled maestro roared. "The boy - you imbeciles! My new student! You strudel loving cretins!!"

The orchestra once more pivoted their attention on the lederhosen child.

"Is he a dwarf?"

"That head is much too large."

"Why is the mother?"

"Ooh, shackles…are we getting shackles?"

The much-exasperated Mahler slapped his forehead. "Ach! I have brought the boy; Albert is his name by the way, to learn by example, your example. A poor example though it may be, but nevertheless today this is the only example I have to hand, and the next idiot to make a stupid comment will be not only playing in Bratislava, they will be living there!"

Much shuffling of feet and twanging of strings followed this.

"Good!" Enthused their director. "Now sit boy, over behind the strings. Ah, yes Frau Einstein, and the chair as well. I do wonder if the police officer is coming back at any time." He then tapped a baton on the podium. "I think a little Night Music. Ach!"

In rapid succession, however, the music stands of the Strings section had fallen over, after the toddler Albert had kicked them with his shoe.

"Very well," simmered the infinitely patient conductor, "let us…"

Albert leant forward and lifted up the toupee of one of violinist with his bow.

Several minutes later, following a screaming fit from the first violinist, Albert was deposited at the back of the orchestra pit, much to the amusement of the brass section.

"Now then … oooh … Gott im Himmel!"

The budding wunderkind started banging the gong with his head, as the gong player stood back in wide appreciation of the boy's talent.

"Put him by the Triangle … he knows that one!" Yelled Mahler.

The prodigy never made the whole distance as he discovered dragging on the sets drapery brought the whole stage down.

"THIS WAY!" And Mahler jumped up and dragged the boy, his chair and his mother out of the orchestra pit, leaving the stage hands to disentangle the ensemble from set for Wagner's - Götterdämmerung.

"Gött und dämmen! Frau Einstein!" yelled the florid Gustav Mahler, "Your child has no ability! No talent! No …don't touch the harp! … fifty marks Frau Einstein - aptitude for music at all!"

Einstein was lying trapped beneath a harp of prodigious proportions, his chair still attached to his leg, staring about the room for all the world a Jack Horner, as if to say 'oh what a good boy am I?'

"But Herr Director?" Frau Einstein attempted.

"But Nothing! I haven't seen such a disaster since Rossini made an opera about slaughterhouses in Albania! Your son will never play! To be a musician requires a gift! From God! Your son is so ungifted – I would call him cursed!"

At which point he dragged them to the front door of the conservatory, child, chair, mother and harp and slammed the door on them.

"Oh meine Liebchen, at least you got a harp out of him."

Descartes and the Valkyrie

Descartes' greatest discovery was Crumpet.

Descartes found himself exhausted from exercising his pineal gland, lay back languidly in his bed and thought of crumpet. After two weeks at sea, he had finally made it to Sweden. First he had be held as a prisoner by the English under the belief he was a spy, the prime reason for this was he spent most of his time undercover in his bed. It was only by the intercession of the French ambassador that he was released with the guarantee he stopped rambling about philosophy. On passing through Holland, the Dutch had impounded his bed after he had failed to declare import duty on a four-poster bed. The Danes kept deporting him back to Holland for several days, as they were worried he had the plague. Now, however, Sweden welcomed him, with the sort of fanfare that gladiators heard when they walked onto the Roman circus.

'Gods, she is ugly' Descartes thought to himself with ambivalence. 'I came all this way, for a walrus? On the other hand, so am I. On the other – other hand, put a moustache on her and she'd look like me! On the other – other – other hand, she's cross-eyed!"

Descartes had a peculiar attraction to cross-eyed women. It was a fascination he could never explain, nor give up, or resist their charm. There was something ineffably poignant, at least to him, of being presented with the face of a woman who not only refused to meet his gaze, but was physically incapable of even looking in his direction.

The city of Stockholm with its myriad of canals, had welcomed Descartes in the same way a conquering hero returned from war. He was carried through the streets held high on his bed, as a procession led with flowers and musicians announced his arrival in the city. People leaned from their windows and tosses sheaves of paper bearing his writings in the street, young girls danced with circlets of flowers arranged in their hair and singing carols. The cannons of the Stockholm castle fired as they approached the castle. All of this was sanctioned by the Queen regnant, Christina Maria Alexandra of Stockholm, regent of all Sweden, Denmark, Finland and Norway of the House of Vasa, or as she liked to be called – Tom.

Queen Christina, or Tom, was robust. Robust in the sense an ox is vigorous, robust in the way a lump of wood is sturdy, and robust in the manner she could punch out a man with a single fist. When she was born there was considerable debate as to whether she was in fact a woman, so hairy and large was the newborn infant. Doctors and midwives were brought in to attest to the newborn's strength and vitality. The infants ability to scream was compared to the dying scream of a troll, and secretly they discussed that the child really was a troll, but the child's mother resolutely declared her baby was a girl and promptly gave her to a wet nurse while her own bruises and scratches were attended to.

The king compounded all of this by a decision that she was to be brought up as a prince of the realm. He had done this, when he was finally forced to realize she was to be an only child, and thus the sole heir to the throne. Accordingly she would need to be politically capable and resourceful as a prince. The fact that she would have probably been so anyway never occurred to him.

At the age of twenty-three, she could ride a horse, throw an axe and wrestle to the ground any man in Stockholm. She had a very strong sense of the importance of physical exercise, and maintained a daily punishing schedule of horse riding, hunting and lifting weights – all of that and she was only five foot tall.

Yet besides her physical appearance she was startlingly witty and could hold her own with any person. Be they bishop, king or logician. Fluent in several languages, including French, Latin, Greek and German, she had a working knowledge of geometry, architecture and most importantly philosophy. So, it was not altogether surprising, that as a queen she might send for Descartes to visit and instruct her in his philosophy. What was surprising was that Descartes did not know of Christina's pronounced attraction towards members of her own gender, or, as she liked to describe them – crumpet.

"Welcome noble philosopher," were her first words to him, quickly followed by, "Now get out of that ridiculous bed and walk with me."

"It's freezing!" protested Descartes. "My lady, I am a philosopher, not a tramp."

"Are you suggesting the Queen of Stockholm, Oslo and the Baltic is a tramp?"

"I'll just get my slippers."

"Take that back to the boat," she ordered a footman, once Descartes had gotten up, as she pointed at the bed.

"The bed? Not the bed? Oh Zeus, father of the gods! Not my bed!"

"Oh, never mind that, we have lots of beds here. Come along. I'll show you to your rooms. How wonderful of you to accept my invitation."

He cast an eye back at his marvellous bed, which was hurried away on the shoulders of guardsmen dressed in blue.

They walked through the corridors of Stockholm castle discussing his journey and his duties as part of the royal household. She was almost a munchkin beside him, barely reaching his shoulder with her head, nevertheless she also had the air of someone was able to command respect from ambassadors, kings and popes, even when peering up from the perspective of an eight year old child. Neither of them could in any way be considered handsome, with their great hooked noses, warts and hunched shoulders. Indeed, from behind it would have been easy to mistake them for father and daughter, so closely did they resemble each other.

"I can not express the joy I feel in your coming all this way," she looked up at him like a cross-eyed child. "Your reputation is spoken of everyone, and I have read all your books. Indeed it would not be too much to say, I am your greatest fan. Now, lessons will begin at five and we shall have breakfast afterwards."

For a moment he looked nonplussed at this, and put his head to one side to see if he had misunderstood; then grinned with such boyish enthusiasm it was surprising his cheeks didn't fall off.

"But of course, the night is the best time for philosophy. We shall work at it all night, as my lady commands."

"Great heavens no, I meant five in the morning. Be at my chambers, punctual at five a.m., you can instruct me in your philosophy while I lift weights."

Descartes almost fainted.

"Five?! … In the morning?" he shrieked, "I've haven't seen that hour since - actually I've never seen that hour. I thought it was a hypothetical construct."

"Ha-ha!" she tapped him on the shoulder, "I do like your sense of humour, and call me Tom. I see we will get along wonderfully."

"Tom?"

"Yes, it's my pet name you see." She smiled. "Here we are, these will be your chambers."

He saw with surprise, a room stark and bare, it seemed to say aloud, herein is cold and discomfort – flee for your lives, and given the reported opulence of the Swedish court, was not all what he had been expecting. The bed itself was little more than a single bed without a headrest, cover or even two pillows, and worst of all lacked a feather quilt.

"I will need my bed," he gasped, as he clung to the door handle, "my proper bed!"

"Nonsense," she tutted, "what you require is exercise and proper diet, and this you shall have in great abundance. This bed is as hard as a board, as are all our beds."

"But, but," he protested, clutching his chest as if his very soul had been struck, "I didn't come all this way for hygienic living and sanitation. Madam! I'm a philosopher!"

"And I am a queen. Five sharp."

At which she clattered off with her entourage to go horse racing.

"She's wearing pants!" Descartes muttered to himself. "And she's cross-eyed! What have I got myself into?"

The next morning at seven, he was woken by his curtains being thrown open and a sergeant of the Swedish guards and his platoon stomping about the room. His night had been awful, his bed little more than a plank with a blanket thrown over it, even his fleas deserted him, so cold and uncomfortable was his sleep.

"Her majesty wishes an audience with the philosopher," the sergeant barked, and Descartes noticed as he opened his eyes the soldier carried a musket.

"Is it five? Already?" He blinked at the windows, never having seen the sun so close to Russia before.

"Her majesty wishes an audience with the philosopher," repeated the sergeant, as he took the musket off his shoulder and fixed a bayonet.

Descartes yawned, dragged his hands across his face, and then felt his chin for stubble.

"Do I have time for a shave?" he sucked his teeth and quizzed the soldier without too much optimism.

The sergeant lowered the musket, cocked its hammer and very quietly growled: "Her majesty wishes an audience with the philosopher."

"I ...?"

"Take hold of the bed!" barked the sergeant

Descartes hid underneath the sheets, as each of the soldiers grabbed a leg of his bed and carried him bodily out the door. The sergeant marching behind them, yelling: "Left! Left! Left! Right! Left!"

The queen was not in her chambers and was not in a good humour.

Descartes found himself tossed out of his bed and carpeted, as she sat at her breakfast. There was a furious silence for several minutes, Descartes knew it was a furious silence by the way she kept snapping her knives and forks in two, and the way it required a steady stream of cutlery from her butler for her to finish the meal.

"Master philosopher, I hope you slept well," she finally turned to him and broke his ordeal of silence.

He decided to bluff his out, and given he had no other strategy it had to be an extraordinary bluff.

"My queen! I have discovered a new theorem! Which I call the Queen Christina Theorem!"

"Oh, be quiet." She sneered. "I do not care if you have discovered a new world and named in my honour. You, master philosopher, have slept in."

"The cockerel was slow," he attempted.

"We don't have cockerels, this is a palace not a barnyard." She tapped her fingers on the table. "Not only do my guards call out the time, but my servants give notice of appointments. No, master philosopher, you are late for class."

"But your majesty, it was not my fault, the bed, it was so hard I was simply unable to sleep."

"Very well, your original bed shall be returned to you, and in the future you will not only attend us at the appointed hour, you shall be early. Is that understood, master philosopher?"

René bowed and his sleeping cap fell off his head and landed on the carpet.

"Only too well, my highness," he muttered.

"Very well, since you have not eaten, would you like some crumpet?" She asked.

"Crumpet?" He stood up and looked perplexed about the balcony, eyeing the servants and guard, "Here? Now?"

She held up a plate.

"Ah, crumpet," René grinned, "but of course."

The morning was a cool summer dawn, which slowly warmed to a conversation about ontology, being and importance of dreaming. They sat on the balcony overlooking the vast gardens and walkways Stockholm castle, a myriad of flowers and trees were arranged in precise order, throughout the royal gardens, tended by waist coated gardeners and spectacled horticulturists. The royal gardens of Sweden were renowned throughout Europe as an astonishing exposition of flowers, mazes, and even a zoo, all the more surprising given the yearly snowstorms that would besiege the grounds requiring a complete removal of the entire garden to glasshouses.

In the background of the balcony a trio of musicians sat waiting, and had been waiting for several hours for a command to play. It was not enough to say they were bored; rather they were bored utterly, with the glazed expression of sheep as they contemplate which tussock of grass to eat next.

"Is that the time already," Christina remarked sarcastically, "I think it's time we went for a ride, don't you?"

"A ride? You mean in a carriage?"

"Heavens no, a horse of course."

"Madam ... I," René looked wistfully at the bed.

"You there!" she pointed at the musicians, "We are going riding. Bring the bed, and bring your instruments."

The musicians picked up Descartes and threw him on the bed, then threw their instruments on top of him. All the while, they swore and whispered in his ear that the first chance they got, he would be growing roses from a reclined position – with his bed. René found himself bouncing about his bed with the enthusiasm of a bag of kittens knowing they are being taken to the lake to be drowned. Desperately he tried bribing the musicians with promises of gold or choirboys, this only served to redouble the rattling and dropping of the bed.

Eventually they came to the royal stables and Descartes was presented with a horse. He was thrown out of his bed and collapsed on the cobbled flooring.

"Your majesty," he wheezed exhausted from holding onto the headboard, "I fear I'm too exhausted to accompany you."

She looked down dismissively and raised an eyebrow.

"Really, master philosopher, this is becoming a burden. Very well, bring the bed while I ride."

Everyone, including Descartes who clutched his chest, looked aghast. She surveyed their faces.

"Very well," she conceded and laughed, "just tie the bed between two horses. I will ride slowly. After all, I don't won't to kill the most famous philosopher in Europe, do I?"

The morning ride turned out to be far more pleasant than René could have hoped for, as they discussed the miracle of birth, the tragedy of death and the splendour of crumpet. In their train the four musicians dragged themselves, now almost thankful to René for providing a luggage mule for their instruments; behind them rode an escort of guards and trailing far behind came Pierre, the long forgotten manservant of René. René, himself, rode at the front feeling like an Eastern king on a Grand Tour. They traipsed first through the royal garden and then out into the vast woodlands that surrounded the castle, until they came to a folly that the queen had not only designed, surveyed, but had also joined in the carpentry and masonry to the amusement of the craftsman. Modelled after the ruin of an Italian tower, it had crumbled perfectly within a few months and was covered with carefully placed moss; it was an ideal setting for philosophy.

"Now master philosopher, let us to lunch."

Christina clapped her hands and from behind the walls of the folly appeared a dozen servants carrying a luncheon table complete with silverware, wine and cold chicken. The musicians picked up their instruments off the top of René and set to a ballet by Jean-Baptiste Lully, this brought forth a group of ballet dancers from behind another ruined wall, who jumped and leaped about the forest ruin in imitation of nymphs and satyrs. The queen was nothing if not a perfectionist.

"My lady, my lady, my lady!" chortled René as he rolled about his bed in delight, "You spoil me so! Oh what a fool was I to have slept in. I promise I shall never do that again. You are indeed a queen of kings."

To this, the queen smiled and cut up his chicken for him. She had realized he was more delicate than she had expected, and she had also decided during the ride that while she was a vigorous twenty-three he was after all fifty-three, that it was best if she did not wish to lose his instruction if she did not burn him out like a candle in an Arctic wind.

He gazed into her cross-eyes, that is to say he attempted to, but found himself deflected by the geometrical impossibility of doing so, it was like trying to have a conversation with a Siamese cat, on the one hand words were exchanged but on the other one had the nagging belief that the cat was thinking about something different altogether. René was, as has been mentioned, fascinated with cross-eyed women, it was not something he could explain, nor a tendency he could ignore, he simply found them irresistible.

"I'll have you know, I rather like crumpet," a little cheekily Christina instructed him, knowing of his pronounced tendencies.

"My Queen," Descartes grinned, instantly warming to this swarthy northern lady and quite forgiving her for the bed incident, "as do we all!"

"Yes," she smiled, hinting at something else altogether, "crumpet is for me is what I desire more than … why, more than anything. Personally, I would prepared to give up my throne just for a piece of crumpet. What say you, master philosopher?"

"Well yes, your majesty," Descartes grinned manically, thinking all his ships had not only come into port, but they had docked themselves and unloaded all the treasures of the Orient. "In fact, it would not be saying too much, to state I came all this way, just on the off chance of a bit of crumpet."

She stopped and looked at him closely.

"I mean," she continued, " and I will put this in no uncertain terms. I do not look upon myself as being crumpet."

"Why yes, I would never … um," René questioned, "what does that make you, besides being queen of all the Scandinavia?"

"That, master philosopher," she smiled and tapped him on the chest, "is a riddle you will have to solve."

"My lady," Descartes almost exploded with anticipation, not

believing his good fortune, "I can't wait to solve the problematic nature of crumpet with you!"

"Ha-ha! For a philosopher, you have much to learn." She grinned mischievously, reading too easily his enthusiasm. "A little wine?"

Descartes with his enormous pineal gland had no head for alcohol and had as much capacity for it as a gnat in a vat of brandy. In the background, he saw his servant Pierre furiously started waving his hands to warn his master, René, however, waved him off so taken with this cross-eyed queen.

"A toast to you my lady," took a sip and fell of the back of the bed and started yelling deliriously "Alright! And here's to Crumpet!"

The transformation was so profound and so instantaneous that even musicians stopped playing and looked over their sheet music to see if he was all right.

"Monsieur? Has he hurt his head?"

The queen came around the bed to see if he was okay, as Pierre hurried up to help his master. Descartes' extraordinary metabolism had taken full control of his mental process and was far ahead of both of them.

"Oh baby!" René finished the glass and yelled. "Look at me! I mean both eyes at the same time!"

This brought the lunch to a frozen silence, the kind of silence that one only hears on glacial ice fields in the middle of winter on the stillest day of the year. After you have not only become lost from the entire artic expedition, you have covered your ears with snow mittens and suddenly realized the white, and very large shape in front of you is a sleeping and pathologically, dangerous Polar bear.

"That will be enough philosophy for this day," Christina frowned, "we shall continue tomorrow at the expected hour."

"But baby!" René cried out as he fell forward on his bed, but it was to no avail as she walked off in a regal huff.

"Monsieur?" Pierre asked.

"Whoa...?"

"Shall we be packing for a return journey, or merely buying a coffin?"

"Ah. Look musicians!"

The next morning he woke with a tremendous headache and a country he didn't know. He also found he was sleeping in his normal bed in the queen's private study. Pierre had taken the liberty of moving him there in the night, knowing full well his master's incapacity to wake before lunch.

The room was an enormous circle of book shelves, as a state library arrayed tier after tier ascending to vast French windows and walkways. Within this, was another circle of seats and tables stacked with volumes bound in leather, manuscripts in a dozen different languages, within this an arc of musical instruments ranging from violas, to harps and pianofortes. In the very centre, lay a couch and a bed. Upon the couch was the queen and in the bed was Descartes.

He awoke with a mouthful of ice water.

"Belah!" was his immediate response, as he coughed out the jug of freezing liquid.

"Good morning, master philosopher."

The queen sat in man's clothing, lifted a dumbbell in one hand, and read Spinoza in the other. To one side Pierre stood fearfully with a pitcher of water and an expression that said, -wake up now, or you won't wake up-. The sun had barely breached the horizon and was slowly coasting across treetops far to the South.

Descartes lay with a frightened wet, expression not unlike the first amphibian to make to dry land.

"Your majesty, I was just dreaming of you."

"A wet dream, I take it?"

"I ... er ... ah ...,"

"Never mind," she grinned. "I see that you have at least managed to be here on time. No doubt in large part to the survival skills of your manservant."

"Yes my factotum, he's rather good, I picked him up at an auction. Ah, where am I?"

"Where you are supposed to be," the Queen frowned, "now then master philosopher. Let us begin."

"But of course," Descartes recovered well, "today's lesson shall be on the purpose of dreams."

"You are a fast one. Well, as to the question of this morning's debate, is it not rather the meaning or the interpretation of dreams?"

"Not at all, your majesty." He smiled. "For who can interpret a vision from another world? Or who can understand the language of a fantasy? Not I. No, the purpose of dreams that is far more important."

"Very well," as she flexed a bicep with the dumbbell Descartes could have sworn he heard the fabric of her shirt tearing. "Continue. You do not mind that I perform my morning exercises do you? I feel that the morning is the most precious time of the day and one should engage in as many activities as one can, before noon. If only to build up an appetite for lunch."

Descartes sat up in his bed and pulled his hair back. He steepled his fingers to his chin and looked at the ceiling. He had discovered this was the best way to have a conversation, if he looked at someone there was that continual distraction of reading their emotions through their eyes - which he found it especially distracting when confronted with a cross-eyed woman.

"The purpose of Dreaming ..." he started but was instantly stopped by the Queen who disliked anyone did not give here their full attention.

"I am here, master philosopher." She said contritely. "Sitting over here. I am not floating above the room; I am not at the other end of a speaking tube. I am here in the same room. Please acknowledge that fact."

He looked at her and this was a fatal mistake, for not only could he not concentrate on espousing his ideas of dreams, he also found it maddeningly confusing to know which eye he was supposed to look at. If he looked at her left eye he felt there was someone standing at his right shoulder, and if considered her right eye he found himself looking at the ceiling to see if there was anything there.

"Go on, I am ready," she said again and then ironically. "Unless of course you require a glass of wine?"

"I ... er ... yes that is to say...," he trailed off, desperately trying to ignore her eyes.

"Dreams, master philosopher, you wish to speak of dreams," she said testily putting down the dumbbell wondering if that was putting him off.

"I ...," but found her wandering eyes so impossible attractive that he froze up altogether.

"Really, master philosopher," she burst out, never being one for other's failings, "given that I have invited you to my home, put you and your servant up, fed and feted you, you might at least have something interesting to say!"

All the while her eyes wandering about the room like a chameleon's stalks.

"I'm sorry," he finally confessed his weakness. "Your majesty, but you are so lovely to look upon, that I am quite overwhelmed."

To this she smiled.

"Ah, I see now why you are so well regarded. Very well, you

may close your eyes on this temple of loveliness, I would not wish to interrupt as great a mind as yours with my angelic form."

Descartes sighed and closed his eyes, he spoke for several minutes and then quite naturally given the time, fell asleep. He woke once more with another face full of ice water.

"Belah!"

"Monsieur," Pierre put down the jug of water and mopped down his owner. "The Queen has left."

"Oh hell." Descartes rolled about the bed. "Was I asleep? Never mind, yes I was, and was she annoyed, again never mind, yes she was. Well, what does she expect? This is the morning!"

"The queen said, she will expect you tomorrow, at the same time."

"What she want's this to happened every morning? Great heavens! This is too much, pack the bed we're off to Paris."

Then dragged a pillow over his head and hid from the world. Pierre lifted the side of the pillow and whispered in his ear.

"Monsieur, she has also posted a guard at the door."

"You mean we're trapped?" René peeked out. "I only came here for the crumpet!"

"I'm afraid so, Monsieur."

"Bribe the guards! Buy a boat! Can you swim?"

"Monsieur, I fear there is no way out, it appears you may actually have to, dare I say it, you may have to work."

"Work?!"

"Yes Monsieur."

"Work?! Great merciful heaven! It's the end of the world!"

"Monsieur, there is one other thing."

"What else could go wrong?"

"You must leave the bed."

"Waaaaah!"

Einstein and the Tsar

In his student years from 1896 to 1900, while his parents were living in Italy, Einstein had gone to Geneva to buy a cuckoo clock. This story is vaguely about this visit, partly about the Russian Tsar, but mostly about clocks.

'Geneva will only be a flying stop.' The Tsar contemplated, languidly stroking his beard with his hand and staring out the window. His children played about him; their spirited laughter and giggles seemed to suit the underlying clatter of the railway. The young Tsarevich had fallen asleep with his head upon the Tsar's leg and the Tsar combed the boy's hair with his fingers and stared out the window.

They were climbing up the hillsides, into the beautiful landscape of the Alps, steep pastures slipped by, dotted with idle cows and industrious Swiss farmers who waved amiably at the train. In return, the young Tsarevich and his entourage would occasionally take pot shots back at them, using one of these new Maxim machine guns with its water cooled barrel, that his uncle, the English Prince George had sent him.

'Yes,' thought the Tsar, 'Europe, unlike the Motherland, is so sophisticated.'

The door to the royal carriage opened, and from the sudden silence of the princesses, the Tsar knew without looking that Grigori Yefimovich Rasputin had entered his carriage. The priest always brought a chill with him that belonged in the northern reaches of Siberia, with his eyes on fire and his roving beard, he should be out somewhere in the white wastes strangling Cossacks and bears. If it were not for the strange hold Rasputin had over the Tsarina, the Tsar would have had him quietly disappeared years ago.

"Good morning, my gracious Tsar." The mad priest muttered

as he took a seat opposite his majesty. "Our Tsarevich is by the grace of our Lady well and uncut this morning."

The Tsar, unsure if this was a greeting, or question, or even a religious pronouncement, merely nodded his head and went on staring out the window.

"My gracious Tsar," the unctuous priest continued, "as we will soon arrive in Geneva, I thought it exigent to raise the matter of the Bishopric in Europe."

'Exigent? Exigent?' thought the Tsar, 'What in damnation did that mean?' and wished he could take the new machine gun to Rasputin.

"It has become obvious from my spies -"

'Has this idiot no subtlety?'

"- that the Metropolitan Pyotr Krutitsky is perhaps planning to destroy the Tsar."

"Really." His majesty finally turned and stared at the intruder. "You mean like the time you suggested all the new born cows in Moscow were the spawn of the Anti-Christ?"

Rasputin shrugged this one off.

"Or what of the time, you prophesied a giant slug was coming to eat the Winter Palace, and had my entire personal regiment sprinkling salt round the royal cabbages."

"They did find a giant slug."

"It was only six inches long."

"Yes! And it would have grown except for my prompt warning!"

"I also seem to remember you suggesting the whole world was coming to an apocalyptic end only last week, when all of Russia would turn mad, destroy itself and my family would be eaten by nuns. Yet, here we are on our summer holidays. How do you answer that one?"

"Well, we did kill the ringleader of the nuns."

"Odd! I thought the mother superior died of fright; you burst in on her at vespers. She was ninety eight years old."

The priest shook the air with his hands. "Yes! And how did she live to such an unholy old age? Surely that was the work of Satan?"

"Monk! Since when was old age a cardinal sin? And, if so, was Methuselah also a spawn of the devil?!"

Rasputin shrugged his shoulders; the long flowing habit barely seemed to only barely contain the massive frame of the mad monk.

"These are difficult questions to answer, my gracious Tsar, perhaps we should answers these another time."

The young Tsarevich choose this moment to wake up, and rubbed the sleep out of his eyes.

"Are there more cows to shoot at, Pater?"

"Not now, my son, we are out of bullets."

At that very moment in the city of Geneva, a bizarre collision of ideologies was taking place high in the roofing of the Russian basilica. The church was the seat of the Russian-Orthodox Metropolitan in Western Europe, built in the traditional Russian mode of six huge gold, gilded, onion-shaped cupolas, it was playing host to a strange conference. For above the Rue Lefort in one of the onions, and in a conversation carried out in rumours, a diabolical plan was taking form.

The Metropolitan, head of the Russian church in the West, declared in a reverential and thoroughly pious voice, "You! You take the bomb, save our glorious Motherland and blow up Rasputin!"

"B... But Metropolitan, surely there is somebody better suited to this work. Dmitri, our Choirmaster Dmitri, he has met Grigori

Yefimovich Rasputin, he studied with him at the seminary?"

The Choirmaster stood up in a sudden flurry of hands, and hit his head on a projecting beam.

"Metropolitan! I have a better idea. Feodor, our young thurifer should carry the bomb in his censer, it will even smoke like a bomb is supposed to smoke! Nobody will expect a young boy to carry a bomb."

"Ah," a thoughtful, young voice echoed in the attic of the church, followed by the sound of young feet running down the wooden steps of the old church and running away into the distance of the street far below, and a young man calling for a taxi.

The aged, and possibly wise, Metropolitan took off his black hat, settled his posterior in the dust of the aged beams and leant his chin upon hands. "Is there nobody to rid us of this terrible curse?" He muttered absently.

The Tsar was coming to visit, all the way from Saint Petersburg. Since, the Tsar was coming, this also meant his family and his entourage were to visit, and that especially meant Rasputin - the Rasputin - would come as well. Rasputin! Rasputin the eater of children and small priests. It could easily mean the end of his Bishopric in Europe, and being sent back in disgrace to Saint Petersburg: he needed a solution and soon.

"Metropole." The astute and terribly capable deacon spoke up in an undertone, an odd combination of inflections, "Perhaps the problem lays not in the act, but in the actor?"

"Are you suggesting we send Sarah Bernhardt?"

"Well, anybody really, but it certainly needs not be one of us. It could be any dupe, a local and a foreigner perhaps. Although, I think someone neither Swiss nor Russian would have the least repercussions."

There was a vague nodding of heads at the wisdom of this

proposal.

"Excellent, so the first idiot that comes along—you mean someone like the Catholic Pope." The Metropolitan rubbed his chin thoughtfully. "Yes, that would be a much better. Very well!" The venerable Metropolitan Pyotr Krutitsky stood up, banged his head on a joist and strode towards the stairs. "Come, we go to meet the Tsar and find an imbecile."

As his retinue quickly filed out behind him, the deacon strayed behind, making a note of the Metropolitan's last comment on his sinister shirt cuff and to pick up the Bishop's hat.

Einstein carefully walked up and down the shop, massive head bobbing sagely, the long dropping moustache covering his pursed lips, his eyes luminous in the dim recesses of the shop, and all around was the sound of Time. Here was a man obsessed with cuckoo clocks, whether they be made in the Black Forest, Switzerland or more hilarious in Furtwangen. Large, small, ornate and simple Einstein found he could never have enough cuckoo in a clock.

Minutes later, he stepped out of the store, carrying an oblong object wrapped in brown paper under his arm, and gazed off into the approaching dusk of the Geneva night. Swallows swam and dived in the crimson light, catching gnats and deranged moths, as the settling sun blazoned itself into the mountains. It filled him with a sense of Swiss good humour and he smiled. Smiling with what he thought to be geniality, at the enchanting landscape, a strange broad grin enveloped by his moustache, a large beam of benevolence radiated from his face, satisfaction shone out of his eyes.

In reality his face was such an abnormal smirk it worried everyone else on the street.

Walking along the shore of Lake Leman, a light spray drifting down 100 yards in air from the Jet d'Eau rained down on him. After a moment's consideration, he carefully placed the oblong parcel underneath his plaid overcoat, worried the mist might damage the delicate mechanism; wrapped, though it was, in two layers of waterproof oiled paper and an extremely sturdy Swiss engineered hardwood casing.

Einstein, however, was taking no chances: a good Swiss cuckoo clock was worth its weight in barometers.

Unbeknownst to him, the moment he stepped from the door of the museum, three plain clothes policemen had started shadowing him, with an intensity bordering on religious zeal. Three good men kept a constant ring of surveillance around the suspect, Albert's every movement was recorded and relayed to the police headquarters: the way he held his head as good Geneva citizen's passed by, the manner of his clothing with its untied shoelaces, and most significantly the disturbing parcel he was attempting to hide under his mackintosh. The fact, that Einstein failed to notice these three identically dressed and heavily built police officers, in sunshades, Homburgs and newspapers, walking in a tight triangle of no more than two meters from his trail, says more about his powers of perception than the incompetence of the Swiss Internal police.

Herr Polizei Inspektor Richter Gunter Mueller, of Geneva's Most Especially Secret Services, or MESS, was convinced they had their man, or in this case: their anarchist. Richter, a man with no pretensions to intelligence, indeed he was often known to ridicule the whole idea of intelligence, was a short, hence often overlooked in promotion, heavily moustached police officer the archetype of secret police. His instructions were to look for a strange looking man, wearing a mackintosh and buying a cuckoo clock.

In reality, the true anarchist was standing on a train station on the outskirts of Geneva, waiting for an opportunity to destroy the entire Russian royal family, and Mueller was chasing the wrong man.

Destiny, however, has a way of bringing people together.

Soon after Einstein was idling at the train station along the Rue du Mont Blanc, surrounded by secret police and Russian monks, and found himself wondering why the city was so busy this evening.

The station in Rue de Lausanne clattered with motion, pigeons exploding out of the cast iron rafters above the rail yards, whole Swiss girls finishing schools poured out of carriages carrying rosaries and Karma Sutras, returning Papal Guards chained smoked cigarettes to make up for the cold turkey they were forced to endure in the Papal City, and trains exhausted from their journeys up the mountains slopes breathed relief, discharging steam and passengers.

With great curiosity, the Russian clergy watched this strange looking man in his plaid suit and wandering eyes. As he stopped and finally realised one of his shoelaces was untied, his pursuing hawks and doves simultaneously stalled into a halt. The tight of priests scrutinized with growing satisfaction this eccentric being. Suddenly there was a flurry of finger pointing, hand waving and head bobbing as they conferred about the package under his arm.

The word 'scapegoat' formed in their collective Jungian sub-consciousness, while words like 'victim', 'loony', 'weirdo' and 'let-flush-his-head-in-toilet' circulated like crows hustling in on the afterbirth of a lamb, as they surrounded him and bodily lifted him into the air, removing the package under his arm, and in that peculiarly Russian manner: all stood two inches from his face and stared straight into his eyes.

"Err, thank you." Einstein chirped, somewhat taken aback at sea of faces crowding his world.

"Are you Swiss?" The Metropolitan asked nonchalantly, fingering the lapel of Einstein's plaid jacket.

"No," Einstein replied, jittery at the closeness of the attention.

"But, I am hoping to change soon." He laughed nervously. "Actually I have only recently lost my German citizenship. I was deported. I am what you may call a stateless person."

"Wonderful!" All the Russians priests clapped in admiration. "So, you are a stateless person?" They all winked at each other. "And your family, they are living in Geneva?"

"No," he returned, basking in their attention. "My parents are working in Padua, in Italy, they own a feather bed factory. Do you sleep well at night?"

The monks all clapped again in appreciation, their black stove pipe hats bobbing back and forth, animated by flapping arms encased in equally black robes. Along with cries of "Oh, you are such a jolly fellow!" - "What sparkling wit!" and "Give him the bomb."

"And—are you catching the train today?" The Metropole asked, his face shining with perspiration.

"Yes," the adolescent said unhappily, "I am taking a clock to my parents. I love to catch trains."

The astute deacon caught Einstein's eye, by moving an inch closer than the rest of the monks. The covey of priests crowded closer, forcing Einstein against a stonewall and making him check his wallet.

"So, your day is free then?" The Metropole almost screamed. "Would you like to catch a couple more trains?"

"Trains?"

"YES—TRAINS!"

"Hmm, a train?"

"YOU KNOW, CHOO-CHOO!"

"Ah yes, a train."

There was much bobbing and nodding of sagacious heads.

"All you have to do is take the train to the next stop, change trains, and return to Geneva on the next returning train. I have a present I wish to give to an old friend of mine."

"Yes take this bombe," the Machiavellian deacon corrected himself, "um, cuckoo clock with you."

"Did you say, cuckoo clock?"

"Yes, we want you to take this clock to the next station and catch the returning train."

"Our friend, Rasputin," all the monks hurriedly crossed themselves, "will be riding on it, and we want to surprise him with our present."

"Ah!" Einstein smiled. "A present."

Einstein arrived at next train stop, where the monks had told him to catch the returning train, carrying the extra package. He fixed his eyes on the shimmering peaks of the Alps. White African cranes were cranking their way up the air stream into Germany passed overhead, the sounds of cow bells clanking on the distance hillsides, where young girls—all called Heidi—sat plaiting their blonde hair with yellow daisies and edelweiss, the tawny thatch roofs of chalets hiding at the base of a enormous valley, and soaring above it all, was an Alpine sky so blue you could have carved it up and used it for icing.

It was the most beautiful of days, yet Einstein found all he could do was think to himself: 'Mein Gott, that guy stinks!'

Einstein found himself next to a heavily bearded, odoriferous stranger, wearing a ferret-trimmed coat, smoking a pipe and clattered nervously on wooden clogs. The stranger was decidedly strange, perhaps it was the way he kept muttering to himself in a wild hebephrenic undertone about how the Proletariat and the workers of the world must unite; or perhaps it was the wild-eyed way he stared manically round the train station, jumping at every noise; or maybe it was simply the final significant detail, the stranger was also carrying a parcel, and it made a delightful ticking noise—not unlike a cuckoo clock.

They stood side by side on a railway platform; Einstein fondling the wrap of his two parcels. The first was his cuckoo clock and the other, the bomb given to him by the Russian priests. The stranger held an identical parcel under his arm, and Einstein found himself drawn again and again to stare at it. The smell of the stranger, however, was overwhelming. Eventually curiosity, Einstein overcame his olfactory senses and he introduced himself.

"Are you catching the train to Geneva, perhaps?"

The stranger swivelled his head aghast at the possibility of conversation.

"Err, no."

"But you are waiting for a train, yes?" Albert replied mildly, trying to move his nose upwind.

"Am I?" muddled the stranger, his eyes rolling about their orbits,
"Um, are you see … I say, aren't these mountains beautiful, all those white cranes and cows."

"Yes, Switzerland is so clean … Err not that … um."

"What?"

"Yes," Albert forced out a strangled reply, "not that it's not clean. No! I mean it's so hygienic, no sorry, that's not what I mean. I say, have you seen the showers… Sorry, I meant flowers over by the toilet… No! Not the toilet, actually the halfwit… No! Ticket office. Aargh! Oh, thank goodness here come the drain … Err, train."

As incongruous—as a line of stretched out revellers at Mardi Gras, dancing, whistling and yelling their way through Saint Peter's Square in the Vatican City—the carriages of the train were rocking from side to side, cha-cha-chaing their way up, under the bemused stare of the mountains.

In a half empty carriage, Albert found himself accompanied by four distinctly odd travellers, odd by virtue of them all being complete opposites and odd because all four sat absolutely rigid in their corners terrified that one of the others would speak.

They comprised, though Albert did know them personally, his holiness Grigori Yefimovich Rasputin with his wild staring eyes and still carrying a gold icon of Saint Basil of the Eccentrics, Herr Polizei Inspektor Richter Gunter Mueller who was looking for the final man. This last fellow was none other than the pungent stranger Einstein had met on the platform: Pyotr Alexeyevich Kropotkin last of the great Russian anarchists.

At that moment, the parcel under the arm of Kropotkin started chiming, the four men stared uneasily at the package and then at each other, this went under for several chimes until there was a distinct sound of cuckoo-cuckoo.

"I hear you have a clock," Albert attempted conversation, which had the effect of a fog rolling in on the school picnic, a palpable feeling of unease spread its way around the carriage.

"What! Who said so?!" nervously Kropotkin retorted.

"But you do," interjected Mueller with disturbing capability, leaning across the carriage and tapping with his finger. "There under your arm, wrapped in brown paper, came the distinct sound of a Swiss Cuckoo Clock, and judging by the precise tone and modulation of the cuck's and the oo's, it has a twenty-four hour clock face with gold lettering and a small blue bird with a yellow beak and red eyes, I would say, it was manufactured in 42 Boulevard Des Tanchees by Monsieur Martin Zwingli, who has a small scar on his upper lips and most probably studied Sanskrit for a hobby."

The other three passengers swivelled their eyes around to this extraordinary font of wisdom.

"Ooooh, this clock, …" attempted Kropotkin.

At that precise moment, the train came to a lurching halt; as the train engineer hauled on the brake to avoid a herd of cattle, lead by the only blind milkmaid in the Alps called Heidi. The carriage exploded into a flurry of tumbling passengers and brown paper parcels, Einstein noticed his world was inverted as he collided with the opposite wall, but this was soon rectified as he fell headfirst on top of Rasputin, who let out a squawk and tried to club his recently inverted attacker into submission with the statue of Saint Basil of the Eccentrics.

The clocks and bombs—of course—became mixed up.

The four men played squirmy on the floor of the carriage amid a great deal of "Excuse me," "Terribly sorry," and "Are you infectious?"

Outside in the carriageway, three nuns fluttered past, tittering at the ménage à quatre with its confused array of clothes, groans and Slavic inflected vowels.

Einstein found himself holding three parcels, each of which made a distinct chiming sound as the rocking of the train started up again, the chimes became progressively louder until there came a peculiarly strangled cry of a cuckoo, which contrary to its customary sound, was more of a "Woch-er-Clunk-Ting!" followed by the thrilling smell of burning cordite.

There was one of those moments where Time seems to stand still, or in this case, all the seconds screeched to a halt, slammed into each, and promptly fell over, as the two Russians leapt for the door, jamming each other in with their bodies, screaming Cossack insults and tearing at each other beards. The police officer pulled out a gun and fired off a round in the ceiling then put it back in his pocket, before trying to wrench one of the parcels out of Einstein's hands, who in turn fell back on the carriage seat, gamely trying to retain ownership of his beloved clocks. Only to find the Russians rejoin the conflict, like a pair of whirling Slavonic wolves diving onto Mueller and Einstein, as they too tried to drag the clocks out of Einstein's hands: presumably so as to throw them out the coach's window.

"It's a bomb! It's a bomb! Don't you understand?" Mueller found himself yelling at the obstinate Einstein. "One of the parcels is a bomb!"

"It's a clock! It's a clock! Don't you understand?" Einstein yelled back, but to no avail as they manhandled him across the carriage and attempted to defenestrate the parcels with him still attached. Held by his legs and thrust out the window, they tried to shake the clocks loose from his steely grip, lurching his torso up and down like a teddy doll, his gangly arms whipping about in the wind. "Next time I go first class!" he screamed to no one in particular, as the train went round the bend.

In the opposite direction another train was coming. It was odd, that of all people who should be on the train, none other than Harry Houdini had chosen to leave Geneva that very afternoon.

Maxwell and the Demon

ITERATION CYCLE ONE

James Clerk Maxwell was a Scotsman. Now if you're a Scotsman this is all you need to know, however for the rest of the world there is much more of a story. The year was 1857 and the rain pattered down from the Edinburgh sky with the endless enthusiasm of a monsoon suffering a cold and wearing a mackintosh. Underneath it all, the department of natural philosophy bore the rain with the air of an old grey aunt who has been left out in the cow paddock to scare the crows away.

Maxwell paused for a moment in his laboratory to stare out the window at the falling rain, and wondered if the modern fascination of vulcanising Wellington boots would ever catch on, the pause was only for a moment as he seemed to suffer from an almost saint vitus dance of animation, leaping from one desk to the next, jumping up on a chair, poking his head in a cupboard and bounding back across the room, with all the vivacity of a rubber haggis being kicked round a golf course.

"Have you heard of this new idea about Duals?" his friend Tait asked him walking in the door without knocking.

Tait was Maxwell's best friend, a bespectacled balding fellow Scotsman, whose idea of a holiday was living in a coffee house in Glasgow reading the latest pamphlets on mathematics.

"We're not fighting with pistols at dawn, if that's what you're thinking," Maxwell replied with a grin and studied an apparatus of glass prisms.

"No," Tait rolled his eyes, "spelled with an A, and not fired with a pistol. Everything has a dual opposite, and I mean everything has a dual opposite."

"Not really a new idea," Maxwell muttered as he rotated the pair of prisms to see if they produced any new types of polarized light. "Greeks had ideas like that, all geometry has ideas like that."

"Ah," Tait lit his pipe, "but not just lines, angles and circles. Even complex things like trees and rabbits have duals."

"Like a negative-rabbit?" Maxwell smirked, before bouncing across the room and peering into a kaleidoscope. "What happens when you add a rabbit to a negative-rabbit ... rabbit stew."

"Not anti-rabbits," Tait smirked back, "but dual-rabbits."

"Duelling rabbits sounds even less convincing than anti-rabbits," Maxwell espoused and sprang back to the polarizing prisms.

"William Rowan Hamilton doesn't seem to think so," Tait said slowly and blew pipe smoke into the air.

Maxwell looked over his glittering prism and squinted at his friend. "Hamilton said that?"

"His very words," Tait produced a paper from his pocket and read aloud, "The principles of duality may very well be applicable to all things, large, abstract and real. Even our very selves may have some counter in another dimension which we can not imagine."

"The Grand Old Man said that?" Maxwell fixed his friend with a thoughtful gaze, and then shook the idea loose, "It's a meaningless statement, for equally I could say there is an exact copy of me in another dimension, only the other one doesn't wear tweed. Preposterous."

Tait tempted his friend by wafting the paper in front of his face. "That's not what the old man said."

"I'm not biting," Maxwell frowned and returned to the prism for a few moments, but his restless energy overcame him and he bounced across the table and grabbed it out of Tait's hand.

"Guess it's salmon season after all," Tait grinned.

Maxwell eyes raced through the paper so fast it was surprising they didn't pop out, and then he tossed it back to Tait with no interest at all before returning to the prism apparatus. "Rubbish, the Grand Old Man has gone a wee bit daftie," he said with his eyes squinting along the axes of the prism. "Has he published this?"

"It's for personal circulation," Tait looked at the paper, "but it makes you think, doesn't it?"

"Everything makes me think - for I am James Clerk Maxwell," dramatically holding his hand up and putting the other in a fob pocket, "now hand me that five iron."

"But what if," Tait folded the paper up and returned it to his pocket, "what if for every action there is an equal and opposite reaction, not just here but in another universe."

Maxwell paused significantly, which in his case was to stand motionless for a full ten seconds, an almost unheard of occurrence in his life.

"What," he said looking puzzled at the desktop, "you mean there's a more general rule of Newton's Third Law?"

"Precisely."

"You know that actually seems to make sense," said Maxwell, "well, as much sense as Glasgow being the cultural centre of the Highlands."

They both folded up with laughter at this, and took several moments to recover.

"All the same," Maxwell continued, "if you really going to generalize, you'd have to include things like books and writing. That surely is preposterous. I mean if you really included everything you'd have to include even the equations we use to understand such abstract ideas."

"Precisely" his friend countered, "Everything has a dual. That's what Hamilton is alluding to."

"He will have to be a little more precise about what a dual is," Maxwell scratched his beard then fell on all fours and began rummaging in a cupboard for an anemometer to test the airspeed in the windless room. "More importantly, has he a mathematics to discuss these abstractions."

"He calls them quaternions," Tait explained.

"Oh that old hat, throw a four numbers together and toss into some complex notation and he thinks he understands the universe." Maxwell paused again as he hung from crossbar on the ceiling. "Still, it is the Grand Old Man. I wonder what happens if I write an equation describing how my dual does things?"

"You mean will that affect your opposite number?" Tait raised an eyebrow hyperbolically.

"Precisely," Maxwell grabbed a bolometer and fell from the ceiling like a cat on to the desk, where he plugged the bolometer onto the end of the anemometer just to see what it would do. "After all, if we are connected to our duals, we should be able to affect their behaviours."

"But might they not affect our actions?" Tait prompted.

"Nonsense we're part of the British Empire," Maxwell retorted, "We won the Austro-Prussian War didn't we?"

"Well no, we skipped that particular war, but go on."

Maxwell grabbed a pencil and began writing out a series of equations in quaternion form, which he thought would match the physics of his dual universe, and at the end of it he tacked on a curious symbol.

"What's the last bit," Tait asked looking over his shoulder.

"Me," Maxwell laughed.

"You?"

"Well, me in the other universe."

It was at this point the World went a bit strange.

#

In another universe, an exact copy of James Clerk Maxwell was in her laboratory, wondering if the modern fascination of vulcanising Wellington boots would ever catch on. This copy of James Clerk Maxwell was exact in ever detail except she wasn't wearing tweed and, of course, she was a woman. Like her opposite James Clerk Maxwell, she was extraordinarily animated and seemed to whirl about the room with all the vivacity of an electrified salmon.

Her friend Peter Guthrie Tait walked in the door without knocking. "Have you heard of this new idea about Duals?"

"Yes, I wrote something down before," she bounced up to the windowsill to adjust a telescope point at the horizon.

"What?" Tait was surprised, "did Hamilton write to you before me?"

"No, I had an idea using quaternions to describe duals. Almost clever, really." She vaulted the desk, grabbed a pantograph to sketch out the diffraction patterns made by the polarizing prisms.

Tait looked at the paper on the desk and whistled.

"That's extraordinary," he said holding the paper up to the light, "that's almost word for word what the Grand Old Woman wrote," he paused and pursed his lips, "except for the bit down the bottom."

"The equations?" James Clerk Maxwell pulled apart a Faraday induction coil to see if the prisms would fit inside it. "Yes, I wondering if I could describe my dual in my dual universe."

"What's the bit at the end?" Tait pulled out his glasses.

"The squiggly bit? I call that my personal constant, it's the bit that describes me, well my other me in the other universe."

"Should it be a function or at the very least a parameter?"

"Does that work?"

"Does for me," Tait insisted, "I would have thought, only don't make it a function of Time, make it a function of our universe."

"Bah," Maxwell snorted and started twiddling with the dials on an electroscope, "might as well say our universe is a parameter of the dual universe. Sorry, Tait it just doesn't make sense."

Tait pulled out pencil and tacked on a function with the dual universe as a parameter, and as he did so the room seemed to shrink a little. Maxwell stopped in her fiddling and looked about the room in some alarm.

"Is it me or have I grown?" she found herself looking down at her friend Tait.

ITERATION CYCLE TWO

In the first universe James Clerk Maxwell looked up at Tait and wondered if arthritis had set in a little early, even for a wee Scotsman.

"Tait is it just me," he asked his friend looking up at him, "or have you grown?"

"Well, five minutes ago, I would have sworn we were the same height," Tait looked down at his friend. "That's odd, it's almost as if you've lopped six inches off your height."

"In under five minutes? Now that's preposterous."

They looked at the heels on Maxwell's shoes, but found beyond their being a little worn they were unchanged. They stood next to the doorframe and measured their heights with a yardstick. Tait was now definitely the taller and unquestionably Maxwell seemed to have diminished by a few inches.

"Perhaps you're ill?" Tait offered.

"What?" Maxwell sounded alarmed, "I'm so dehydrated I'm shrinking?"

"It's possible," Tait shrugged.

"No it isn't, if I was that dehydrated I would grow thinner not shorter. No, there has to be another explanation."

"Well let's work backwards, what has happened in the last quarter of an hour since I arrived."

"Ha!" Maxwell laughed as he pointed at all the scientific apparatus he had dragged out of the cupboards and tacked onto each other, in his pursuit of examining polarizing light. "There may be a few things."

They went through each of the instruments and examined them carefully, but found nothing that could even begin to explain how he might have lost his height.

"My wife is going to be so annoyed," he sighed.

"Maybe it's time to see the physician?" Tait prompted, "This could be serious."

"Serious would be turning up home and asking the wife to get the marmalade off the top shelf. No, we are not leaving here before I've found an explanation."

They looked about the room for anything at all that might help with the problem. Tait stood in front of the workbench and stared at the equations on the piece of paper as Maxwell ran up a ladder to the rafters.

"James?" he quizzed, as Maxwell's posterior was about to disappear into the ceiling. "I don't remember this symbol being a function."

Maxwell fell out of the ceiling and landed on the desk.

"What?" he said with more surprise than Queen Victoria discovering the Loch Ness Monster in her bathtub. "I didn't write that. Did you?"

"No, it's all yours."

"But that would make the equation of the dual universe a function of this universe, or a function of me to be precise."

"If I wasn't brilliant I'd be getting a head ache at this point," Tait grinned. "Mind you, the equation really was the last thing you did before you shrank."

They looked at each other in disbelief.

"Are you thinking what I'm thinking?" asked Maxwell.

"Wouldn't that make me the dual?" Tait laughed then frowned when he saw how annoyed Maxwell looked. "I'm thinking, the dual

Maxwell wrote in that function, and since every change in one universe must have an equal and opposite reflection in the other. Therefore it's responsible for your transformation."

"It's not quite Newton," Maxwell made a face that invented a whole new topography, "but it does have the gist of making sense. Here, let's rewrite the equation, but leave out the bit which refers to me."

He drew a pencil line through the first one and wrote a series of equations, only this time without himself as a function. Suddenly he found himself looking eye to eye with Tait.

"Great Robbie Burns!" he yelled, "it worked!"

"It's like magic," Tait cried with joy.

"It is magic!" Maxwell whooped.

They stared in amazement at the paper.

"I wonder what else we could do?" Tait murmured.

Maxwell began walking up and down the aisle with his hands behind his back like a wound-up mechanical toy soldier. In his head he was formulating a dozen different ways he could manipulate the equations to solve various physical problems he had been working on, then in a flash he leapt upon the apparatus of polarizing prisms and started measuring their dimensions and refractive index.

"What are you thinking?" Tait asked almost dancing with joy. "What's next?"

"What if we include this machine into the equation? What if …?"

"That's brilliant!"

#

In the Dual Universe, there was a brilliant flash of rainbow light as everywhere the colours of the rainbow cascaded through windows, gems, clouds and water. The whole Universe, as through every galaxy the index of refraction of all light shifted by a negligible fraction. None of which was noticeable by any of the inhabitants, as they blinded by the rainbows in their eyes.

"Ouch!" Maxwell and her friend Tait exclaimed, as their world became an unbearable wash of colours, the very lenses in their eyes filling with a kaleidoscope of images and colours.

"What happened?" asked Maxwell.

"I can't see properly!" Tait shot back at her.

"I'm blind too! Are we having strokes?"

"What simultaneously? Both of us? At our ages?"

"It must be gas!" Maxwell cried and dragged her friend to the door. "Quick outside before we collapse."

It was to no avail, as outside there was an equally bewildering rainbow of images filling their visual field and making it all but impossible to make sense of anything. Every time they leant out to grab something they were deceived by the multiplicity of images and they ended up flailing at empty space.

"Wait, I remember I don't have gas."

"But what is happening?!" Tait became frantic.

"It's the index of refraction in our eyes," Maxwell gasped for breath and felt nauseous. "It's been changed - somehow."

"Will we go blind?"

Maxwell threw up her breakfast and held onto the door.

"I don't think so, I also don't believe it's gas," she explained between retches, "quick back inside. I have an idea."

They staggered once more into the laboratory and for many anxious minutes as Maxwell rustled through the cupboards and drawers, before she pulled out a sheet of coloured glass and held it to Tait's face.

"There," she said and then held another piece up to her own face.

"Green glass?" Tait was relieved and astonished at the same

time.

"Monochrome light," Maxwell explained to her friend, "only one species of light enters our eyes, so no refractive dispersion and no rainbows everywhere. Just keep your eyes closed while I make some masks to hold it to our faces."

Several minutes later she had made some masks out of a pair of stereoscopes and tied them to their faces. It was as peculiar a sight as one was ever likely to witness in the heady days of the Victorian Industrial revolution.

"What going on?" Tait asked her, feeling his stomach revert to normal.

"I don't know," Maxwell put her head outside, "but we're not the only ones."

Up and down the street, people were thrashing about with their arms, crying in bewilderment or staring up into the sky with amazement.

"What could have caused this?" Tait asked.

Maxwell thought very carefully for a full minute.

"We may have, well, indirectly we may have," she stared through the mask like a drunken parrot at her friend. "The Equation."

"Nonsense," Tait retorted, "a mere equation can't change the whole world."

"No, but it might change a whole other world."

"The Dual World."

"Precisely, if for every action in this world …"

"…there is an equal and opposite reaction in the dual world."

"Then my Dual Maxwell has inadvertently changed this world."

"So what do we do?" Tait asked.

"We have to put right what is wrong," Dual Maxwell grabbed

her pencil and began scribbling equations. "Obviously my opposite number has made an adjustment to the refractive index of this world, presumably because said opposite number had been working with polarizing prisms. Ergo, let's send a message."

"But how?"

"We need to show our opponent the terrible changes a few parameters can make," Dual Maxwell leapt on the anemometer, the device used to measure wind velocity and held a ruler to it, "Wind velocity is dependent on air-viscosity, let's add that to the equation and nudge it all so little to the left and see what happens."

She pounced on the paper and after a great deal of scribbling managed to shift the room's airspeed by an imaginary fraction.

"There," Dual Maxwell breathed deeply, "just a little and just enough, that should make them understand."

ITERATION CYCLE THREE

A force-twelve gale was blowing through the Maxwell's laboratory. Maxwell and his friend were hanging on by their fingertips to the desk and screaming in terror.

Tait yelled above the roar, "I know Scottish weather can be bad but did we forget to close the window?"

Maxwell dragged himself to the window and slowly pushed it shut, only to find the wind was still whirling around the room with all the energy of an angry maelstrom.

"It's in the room!" Maxwell let go of the wind and flew across to Tait. They collided and together flew out the front door with the alacrity of two lemmings learning to fly off Edinburgh castle. They landed in the garden outside and were rained upon by the various scientific instruments Maxwell had been using.

"I heard of storm in a teacup," said Maxwell as the bolometer landed in his lap, "but I've always looked upon it metaphorically not meteorologically."

"What happened? Was that a bomb?" Tait asked staggering to his feet.

"If it's a bomb," Maxwell rubbed his head as a microscope landed, "it's still going off. No, I don't think we need look too far for an explanation, for I expect my Dual Maxwell has been doing just the same as I have, and fiddling with the laws of the universe."

"Wait - that could kill us!" Tait's terror escalated to horror.

"Yes, so we need to send them a message and quickly."

"Like stop what you're doing, it's murder!"

"Something along those lines," Maxwell grinned weakly, "now we need to find that paper with the equations on it before all of Scotland disappears into Loch Ness."

They scrambled about the lawn picking scientific instruments and paraphernalia out of the dahlias, until they found the equation paper jammed on an iron railing, all the while the roar of the storm in the laboratory thundered on regardless.

"Write out what you wrote before!" Tait yelled.

"Why?"

"Maybe that's the bit that's causing the storm!"

"But that was optics, that storm in there is meteorological!"

"Meteorological or not, I nearly got decapitated by volume F of Encyclopaedia Britannica! Now put everything back to normal."

Maxwell scribbled out the offending equation but nothing happened, they watched the storm growing in the laboratory.

"I think it's getting worse," Maxwell frowned in desperation. "Maybe if we sent our Duals a message."

"What?" Tait looked amazed at his friend, "Add something to the equation like a coded message?"

"Yes, if we add something to the equation that changes their world just enough to make them realize that what they are doing is violently affecting our world."

"But what?"

"We need to shock them, just a little, just enough to get their attention," Maxwell grabbed the Faraday induction coil out of a bush and measured the number of coils on the wire, "this should do nicely."

#

The bewildering kaleidoscope of images ceased in Dual Maxwell's world.

"It worked!" said Dual Tait and he gingerly removed the filtered glass from his face. "I can see!"

"So can I," said Maxwell as she stopped feeling nauseous, "I think we got through to them. Alright whatever we do, we mustn't make any more changes…"

"Wah!" They both screamed as a mild shot of electricity raced around the room, and this time she really did leap about like an electrified salmon.

"What was that?" Tait asked in surprise.

"Was it the physical laws settling down?" Maxwell asked equally shocked. "I certainly hope that doesn't…Wah!"

They jumped up and collapsed on the floor.

"Was that you?" Tait asked.

"No, all I can suggest is my Dual Maxwell is still tampering with…Wah!"

They ran out of the room in alarm.

"I think we're safe," Maxwell finished. "It seems to be confined to the laboratory."

"You think they may be reworking the equations?" Tait asked staring back in the room.

"Great Robbie Burns! I hope not," said Maxwell, "but wait, the refractive index for everything is back to normal. That means our counterparts first reverted the equations."

"So why are they now shocking us?"

"Either they don't understand what's happening and they are simply writing random equations to see where they lead, or they are attacking us by changing the laws of our universe."

"Why on earth for?" Tait shook his head in disbelief.

"Perhaps they believe we are attacking their universe, after all, I did start changing the equations first," Dual Maxwell walked back and forth in her skirts. "It is possible of course, well at this point anything is possible."

"Wah!" they both jumped on the spot.

"The area of effect is increasing!" Dual Maxwell exclaimed.

"Wah!" they both shuddered as the electric shock made them tense their muscles.

"If it grows … wah! It might eventually cover the …wah!" Tait tried to say.

"Back in the … wah!" Dual Maxwell yelled, as they were shocked yet again. "We have to insulate our … wah!"

They crawled back into the room, leaping every few seconds into the air, until Dual Maxwell grimacing her teeth was able to grab the pencil and scribble out the world equation with factors for the anemometer and the air wind velocity crossed out.

Nothing happened and the shocking became more intense.

"Add something … wah! Else," Tait rolled his eyes in pain, "anything … wah!"

"Can't think … wah!" Dual Maxwell yelled.

Tait pointed at a set of weighing scales. "Try gravity! … Wah!"

"Can't … wah! Too dangerous … wah!"

"Just do it! … wah!"

ITERATION CYCLE FOUR

In the first universe the wind storm in the laboratory ceased.

"We got through to the other side!" Maxwell yelped as the rush of wind fell to a peculiar silence. "We're communicating!"

"Try sending a coded message, like the telegraph." Tait suggested as they went back inside.

"I don't know telegraphic messaging," Maxwell explained. "And how do I put that into an equation?"

"Try a exponential function, anyone will do, just as long as signal grows until they signal back."

Maxwell tacked an exponential function around the electrical component and stood back. They waited for a few minutes and nothing happened.

"Maybe they've had enough sense to stop?" Tait suggested, "maybe we should stop?"

"Yes, I expect your right," Maxwell nodded his head and scribbled out completely the electrical component.

"All right," said Tait and they grinned at each other in relief, "it's over."

At that moment they started to levitate off the ground, Maxwell grabbed the desk and Tait grabbed onto Maxwell's legs.

"Not Gravity!" Maxwell shrieked, "That'll be the end of us all!"

About the room bits of smashed furniture, broken glass wear, pencils, paper, scientific instruments floated free from the floor in a strange aerial mound about the desk.

"Look!" Maxwell pointed, "It's localised! The further from this desk the less the effect! I've got an idea."

"No!" Tait yelled.

Maxwell let go of the desk and they gently floated to the

ceiling to find themselves resting upside above the room.

"It's alright," Maxwell said, and breathed slowly, "that point on the desk must be where Dual Maxwell is writing the equations. That's where the effect is the greatest; it's not as bad as I feared."

"I feel ill," Tait looked quite green hanging upside down in mid-air.

"Must be the world's first case of airsickness," Maxwell grinned. "Think Tait old man, we are the first people in the history of the world to fly if not the history of the universe!"

"I think I'm going to be sick," Tait moaned.

"Here use this paper bag," Maxwell said inadvertently inventing the world's first airsickness bag.

"Now what?" Tait said after he emptied his stomach.

"Lets kick off the ceiling," Maxwell grinned, "if I'm right, the field of gravity should increase towards the door and we should naturally drift to the ground."

This happened exactly as he predicted and they found themselves in a crumpled heap by the door.

"Great Guns!" Maxwell exclaimed, "we're discovered the world's first anti-gravity field, and we're the first people to fly and invent air traffic control."

"And proved the existence of another universe," Tait sighed, "Although I wished I didn't feel so poorly."

"It's for science!" Maxwell pointed a finger at the ceiling, "This is as great as Pythagoras proving his theorem of triangles, Columbus discovering the New World. Hmm, I wonder what else we can do?"

"No James," Tait shook his head in alarm, "enough is enough. Besides have you asked yourself, why our Duals are playing around with our gravity?"

"To tell us the electricity equation worked!"

"On the other hand, didn't the whirlwind almost kill us? And what if the shrinking had gone on? You might not have been able to pick up the pencil."

"Technical trivialities," Maxwell said and leapt into the air across the arc of gravity to the other side of the room. "Woah! Tait we have to try something else!"

"We could destroy a whole universe with this meddling! We could destroy our universe!"

"We won't know until we try."

"Remember Icarus."

"Icarus flew too close to the Sun," Maxwell dragged the paper out of his pocket before his friend could stop him, "now what else could we try? A steam engine?"

"James! No!" Tait wrung his hands together. "Haven't you learned enough already?"

"It's for science, for science we can never learn enough!" Maxwell dragged a model steam engine from out of a cupboard. "Hmm, motive power?"

#

In the Dual Universe the electrical shocks stopped and the two other heroes breathed sighs of relief. The room was a shambles, the two heroes were covered in sweat and a sense their lives would never be the same.

"Is it over?" asked Tait, viscerally shaking from the repeated electric shocks.

"I think so," said Dual Maxwell, "I really should remove the gravity function."

"What if they use it?" Tait asked wiping the sweat from his eyes.

"You mean – the other me?" Dual Maxwell looked curiously at her friend, "I doubt it, once I knew what it was capable of doing I would never do such a thing."

"What if they don't know?" Tait panted heavily, "What if they are simply writing equations to see what would happen? I say – is it me, or is it rather hot in here?"

"That's probably the air-conditioning."

"We live in Scotland, there is no air-conditioning."

"Outside!" Dual Maxwell yelled and grabbed the paper with equations on it.

As they ran out the door, there was a sudden whoosh of air and they were expelled like a couple of corks from a cider bottle. They landed so heavily in the shrubbery that Dual Maxwell was surprised neither of them was injured.

"If this keeps up," said Tait brushing himself down, "we will be turned into Englishmen, and I can't think of a worse fate."

"I'm a woman," Dual Maxwell looked dourly at her friend, "to be turned into a man is the worse fate."

Still hanging onto the paper she flattened it against the stone fence and scratched out the terms involving gravity.

"Perhaps they still don't understand the problems we are having in our universe?" Tait suggested.

"I've got it," Dual Maxwell exultantly held a finger up, "we will do the exact same thing to them as they are doing to us, that way they're bound to know what's happening. Trouble is - what precisely are they doing?"

"Something to do with heat and air?" Tait suggested and they looked at each other with the same thought.

"Steam Engine!" they yelled together.

"Quick, go back in and grab the model engine," Dual Maxwell said to Tait and pushed him towards the door, as a great wave of heat from a furnace blew through the portal.

"You go in!" Tait yelled back, "It's your equation!"

"Yes, it is my equation," Dual Maxwell grinned at her friend, "and if I remember correctly, it was you who turned my constant into a function and made all this mess happen in the first place."

"Ah," Tait made a wry smile and slapped himself in the forehead, "I'll give you that, but if I don't come out again make sure my golf clubs go to a good home."

He took a couple of deep breaths and ran in the room, all the while the rush of air entering and leaving the room began to take on the sound of a giant bellows. After a few tense moments there was a muffled cry and Tait shot out of the room so fast he somersaulted like a highland caber and knocked Dual Maxwell to the ground.

"Got it!" Tait stood up covered in sweat and handed Dual Maxwell the tiny model steam engine, "I could do with a glass of water, no wait make that whiskey I'm still Scottish," he felt in his trouser pocket, "yes, still Scottish."

Dual Maxwell rolled her eyes, and then looked carefully at the notation on the model engine.

"Got it, ten psi at 212 degrees Fahrenheit," she wrote on the equation paper muttering to her self, "now we add in the Carnot cycle in some vague way."

\#

"Great Ghost of Robbie Burns!" the first Maxwell cried, as he broke out in a sweat, "is it just me or did the monsoon just come to Edinburgh?"

Then they watched in relief as all the objects in the room fell to the ground.

"Gravity is normal again," Tait looked about the laboratory mystified, "James, is that steam or fog forming in the room? I mean shouldn't the Dual Universe have the steam in it, not ours?"

Maxwell thought very carefully for several moments.

"What if our Duals are also using a steam engine in the personal function?" Maxwell mused as sweat poured out of his eyes.

"We must stop!" Tait cried and tried to drag the pencil out of Maxwell's hand. "Before it's too late!"

"It could also be too late to stop!" Maxwell broke the pencil dragging it out of Tait hands. "We must go on! Don't you see, now they are doing the exact same thing as we are, we need to balance the equations to put everything back to normal!"

"Madness! Madness!" Tait dragged at his hair with his fingers, "I say - it is hot."

There was a sudden rush of hot air and they found themselves gasping for breath.

"It's like an oven in here," said Maxwell.

They looked at each other in alarm and jumped out the windows, just as an enormous rush of hot air pushed them out like a piston.

"I fear you may be right," Maxwell said ruefully, still holding onto the piece of paper. "We need to stop them, before they completely stop our universe."

"Easy to say," Tait breathed a sigh of relief, as his friend finally understood the danger they were in, "but how?"

"We need to put some sort of shield or insulation around the other universe, we need to control what they do to our universe."

"A sort of infinite wall constant?" Tait pursed his lips in thought, "It won't work. Their universe is connected to our universe, if we cut it off then it could just very well backfire onto us."

"How about a door function in the wall? One that we control, and our Duals can't touch?"

"Like a box around the other universe?"

"Precisely!"

"James, it's not some demon in a box we're dealing with here. It's a whole other universe."

"It's them or us, and I'd prefer it's them," Maxwell clenched his hands in frustration.

With this, Maxwell tacked a wall function onto the Universal Equation, which meant only he was able to write equations and not his Dual, it was the penultimate act in the great battle. Thereupon Edinburgh returned to its unending drizzle, as the people were left wondering if the modern fascination of vulcanising Wellington boots would ever catch on.

ITERATION CYCLE FIVE

The author grinned but shook his head with annoyance at the altogether too brief story he had written about James Clerk Maxwell and his friend Peter Guthrie Tait. He drew several lines across the front page and threw it in the waste bin.

"No one would ever read that nonsense."

This resulted in an entire universe imploding, but the author was never to know that, and the space you are in … just became ever so smaller.

Einstein and the Alps

Do not climb every mountain,
it's a bloody stupid song – A. Einstein's song book

Einstein went a walking high in the mountains of the Austrian Alps. Tier after tier of mountain peaks rose before him, distant and intangible from afar, they hung like tapestries in god's immense hall, crested with dazzling ice and carpeted with green moss, they painted a history of the world far removed. This ancient vista had dripped away molecule by molecule, over aeons so vast that humanity had no meaning beyond a mention in the lost and found of the universal press. On his days away from the patent office, Albert would go there to contemplate the mysteries of the universe, pondering what lay beneath the thin fabric of reality, to question how the clockwork of the cosmos was tidily put together – although the most important reason was that he got away from his annoying landlady, who insisted on his playing cribbage with them whenever he came home.

"Albert! Come in!" Cried his petite landlady, as he passed her door, upon returning from the office.

"Ach, I must just ..," he would attempt.

"Ach - nothing, come in, come in, I have cake and coffee," Frau Sauerkopf would exclaim, as he tried edging away. Einstein had noticed she always had cake and coffee; it wasn't so much a statement as a universal law, as set down on embroidered napkins in her tiny flat.

"Really, Frau, I must ..."

She would first outmanoeuvre him by darting to his left, and then to his right, this Vienna waltz continued until she had managed to place him between the door and herself, and then proceeded to push him in, regardless of his protestations, until he was sitting at the kitchen table with a coffee in hand and a cake - half in his mouth. It was pointless to argue that he needed time to think about something, as she would point out that playing cribbage required no thinking at all, to this he would say that of course cribbage required some thought, and her answer was that the way he played was with a complete absence of thought. To which he had no reply, as it was such a boring game he never gave it any thought at all. This delighted Frau Sauerkopf, as it meant she never lost a game to him, and that delighted her even more, as each game brought a small bet, which she inevitably won.

So it was to the remote climes of the Alps, Einstein would spend his weekends, staring at the heavens wondering if he could find a way to triumph at cribbage. A chill wind rolled off the mountains, bringing a hint of glacier and a thrill of icy rain to the highlands clime. He had caught a train from Bern, all the way across the border to Austria just to see what was out there, and discovered to him amazement, that it also contained mountains, not unlike the ones he had left behind, and it led him to wondering if there was any limit to the number mountains you could have before there was a continental slide into the ocean. The mountains were indifferent to this geographical question, and continued on with their splendour.

He met a mountain guide that morning, just as he left the train, and they discussed various walking routes.

"Ja, I would like to climb every mountain."

"You would have to ford every stream."

"And follow every rainbow."

"Till you ran out of steam."

In the original Austrian it sounded a lot more lyrical, yet the guide was able to offer him some useful advice like - under no circumstances should he go, as the weather was sure to close in after lunch and he would surely die of exposure if he was foolhardy enough to leave the town. Half an hour later, Einstein was up in the mountains gazing fondly at the marvellous sight of the sun streaming down through brilliant clouds, which drifted across the knife edges of the peaks.

"Ach, mein gott, so fantastic," as the weather closed in behind him.

Before long, he was stumbling in the murk, barely able to see the path beneath his feet, as he strove to find his way back to the town from where he had started.

In the opposite direction, and to Einstein's good fortune, was headed another explorer of a more serious nature, who with compass in hand and sure of foot - Sir Edward Whymper was as formidable as the mountains he climbed. He had lead expeditions from Iceland to Tierra del Feugo, deep into the highlands of New Guinea, across the wastes of Siberia up into the uncharted peaks of the Altai Ranges. His exploits were a source of inspiration to legions of adventure writers, and sped the blood of adolescents across the globe. Sir Edward was a colossus of man who at that moment was on his way to achieve yet another milestone in his career, to climb the forbidding heights of Mount Großglockner.

"Hello good sir, out for a jaunt?" The enthusiastic Englishman grinned as he found Einstein stumbling about on the track.

"Ach, why yes, they said I would get lost, but I have found this path."

Sir Whymper looked down at the path in puzzlement, "Was not the path here already?"

"Ja, but my mountain skills are highly developed," joked Einstein, knowing full well he was completely lost.

"Ah, that's what I wanted to hear." Whymper misunderstood Einstein and believed him to be a genuine mountain guide. "Well, I was on my way to find a guide from the next village, perhaps you could guide me?"

Einstein, who understanding of English was not perfect at this point in his life, was only too delighted to help out this courteous gentlemen, "Ach, of course, this would be fun. I often come to the mountains." Never stopping to consider the dangers entailed.

"Excellent, I will pay five pfennigs per hour, and afterwards I will even buy you dinner at the tavern."

At this point, Einstein threw all caution to the wind and agreed whole-heartedly. Fortunately for them, there was a short break in the clouds and a vague route could be discerned to the base of Mount Großglockner. So it was more of an excited Sir Whymper leading a similarly animated Einstein as they tramped their way to the glacier, at the base of what is considered one of Europe's most dangerous mountains. They passed the short time till lunch swapping songs, and jumping over alpine rivulets, or discussed the various petite mountain flowers and insects that flourished in the warm spring air, Whymper who considered himself something of expert on the botany and entomology, would offer up scraps of esoteric knowledge, to which Einstein would nod reassuringly, only serving to reinforce Whymper's mistaken belief in Einstein's abilities as a guide. They continued this delightful wandering, until lunchtime when they found themselves at the moraine of a huge glacier that lay shattered at the base of the soaring mountain.

It was also here, that they were fortunate enough to meet a genuine guide leading another walking party. There were the ritual friendly overtures that strangers exchange upon meeting in an isolated place, and it was not until the genuine guide came to asking Whymper and Einstein about their quest was there any hint of disquiet.

"Oh, we're off to climb that beast!" Whymper fairly trumpeted to the world.

"You intend to climb, Mount Großglockner?" The guide asked in perplexity. "Yes, I have climbed it many times, but you can not climb from here, for the glacier is impassable. It is as death trap."

"Nonsense," retorted Whymper, "my guide assures me we can easily do this. He has often climbed these regions."

Einstein who was pleasantly off in his own world, smiled in accompaniment, as if to say, yes he knew the routes in the same way a blind rabbit appreciates the dangers of a steel trap, not realizing how deadly serious was the moment. The real guide looked at Einstein, sizing him up. It was the complacent, dreamy look in Einstein's eyes that put him off; at the same time he did not wish to seem timid before his own party, so in the end he wished them luck and continued on their way.

"You are a jolly fellow," Whymper patted Einstein on the back as they smoked their after lunch pipes, "now, where shall we begin?"

Einstein who at that moment was considering an equation, pointed with his pipe stem, in the general direction of the mountain in front of them. "It's a problem, no question about that."

"Yes, it is," smiled Whymper, waiting expectantly, having all the confidence of the man who been hoodwinked into buying all of France from someone called Murphy.

At that moment, the solution to the equation came to Einstein fitting neatly in place inside his minds eye, and then to no one in particular he said, "There. Ah."

"There?" Whymper looked at the crooked path that Einstein seemingly alluded to. "There?"

"Hmm, what, oh yes."

A broken line of crevasses littered the path, blue ice daggers reached out from torn fissures of the ice. Whymper walked up to the ice, and drop a pebble into the gloomy depths. It chinked its way down through the ice into silence and vanished.

"This way, you're sure it this way?"

As ever, Einstein his mind on an imaginary space, "Oh yes, yes, I think so."

"Well, good to hear," Whymper patted Einstein on the shoulder and they walked out on the ice. They're hobnail boots sufficed to grip the ice, and equipped with walking sticks and innate ineptness that would have impressed the captain of the Titanic, they soon made their way across the ice until they came to a crevasse too wide to jump across.

"What now, what do you think?" The English lord prompted.

"Hmm, electrons do jump from one energy level to the next. Yes."

"They do? Well then, we shall too, by Jove!"

He walked back, considered the distance across the hole, and with a loud, "Huzzah!" charged forward and leaped across the chasm.

"I say, this is fun, come on, this won't take long at all."

Einstein, his mind half on the situation followed in train, leaping across the void, only distantly aware of the near death experience. They continued this exhilarating jumping from level to level up the glacier, sometimes near falling, sometimes tottering on disaster, always managing to just make the next slab of ice. Indeed all seemed to be going their way, when in the midst of their perambulations as Einstein landed next to Whymper, there was remarkable crash behind them and they turned to see the block of ice they had been just standing on, break up and disappear into a blue void.

"Good Lord! That was a near thing," Whymper remarked merrily.

"Yes," replied Einstein, and for the first time that day, a hint unease entered his voice. "We seem to be stuck."

"Stuck? Stuck?" Piped Whymper, as he looked where Einstein was staring.

In front of them they saw yet another crevasse that was so large and encompassing that they had become marooned on an island block of ice.

"I see. Nasty. Can't go back, and no way forward. It's not so much a dilemma as a moment of incalculable terror. I love it!" Whymper's enthusiasm for danger knew no bounds, and it often resulted in him being escorted from restaurants for fire breathing with the brandy. "Well, then, spot of bother this, no denying. So, what do we do now?"

Einstein looked him, and that feeling of unease grew into mild alarm.

"Pipe."

"Oh, yes, time for a pipe," as they sat down shoulder to shoulder and considered the possibilities.

It had finally begun to dawn on Einstein that he was, to use the manner of the English man, in a spot of bother. He looked up the mountain, and then back down again, an unending array of jagged crevasses and broken monoliths of ice disappeared before them in the gathering gloom. He saw the clouds coming down and with them a corresponding change in both temperature and the cracking sounds of the glacier they were riding. A little drizzle started forcing them to cover their pipes with their hands to stop them going out, this soon became a downpour, which discharged into the crevasses and unbeknownst to them, to lubricate the giants blocks of ice and they felt the monolith of ice creak beneath them.

"Ah ..." began Whymper, this became, "... aaaaah!"

The huge block of ice slowly toppled back and they slid off the top and fell arms cambering into the blue void beneath, bouncing off the sides of the crevasse in terror they fell with a splash into a river underneath the glacier, the freezing cold cut straight through their clothes, as they floundered in the gloom until they dragged themselves onto a large rock out of the water.

Above them a vault of ice lunged upwards and they could make out a narrow window of sky far above them, there was obviously little chance of their climbing out, yet this left Whymper unfazed.

"Bother," said Whymper, "my pipes gone out. Right then, what next?"

Einstein looked at him in amazement, and held his hands up to express his sense of powerlessness. Whymper looked at the hands and somehow caught the idea Einstein meant they should dig their ways out.

"But we don't have a shovel," he pointed out, then out of the corner of his eye he saw a narrow ice tunnel that lead off, "right you are, you mean go through the ice tunnels, I say you are a clever fellow."

Einstein gibbered with cold and fear. Through the icy waters they splashed their way illuminated only by a sapphire light and Whymper's unquenchable enthusiasm.

"This reminds me of the time I was lost in the Himalayas. Almost had to eat a Sherpa. Then the dreadful fellow fell off a cliff before I explained I was only joking. Sherpas aren't what they used to be, let me tell you."

Very soon they realized they were in a maze of towering ice walls, with endless dead ends and icy pitfalls, with each moment the cold water stealing their warmth and sapping their energy, they might have continued onto till their doom, had they not suddenly fallen into an even deeper crevasse and landed into a swiftly running stream that swept beneath the glacier, until with a rush they shot out from under the glacier. Dragging each other from the freezing waters they collapsed on a stony bank of a stream, almost exactly where they had started their trek across the broken ice shelf.

"Great Scott!" Whymper shuddered, "We made it! What an adventure!" They stamped their feet and slapped their arms to warm themselves. "Truly astounding, this will be my best story for the Imperial Geographical Society – why ever! I would have thought. By Jove! You are a good guide!"

A teeth chattering Einstein grimaced.

Whymper who was also a scout master for Wolverhampton quickly set about making a fire, about which they sat and drank Whymper's whiskey, who effusively congratulated Einstein on his glacial expertise, his speleology, and his all round sportsmanship. Einstein just stared in a daze at the Englishman.

"We made it," the Englishman said, as the conversation waned in his favour, "well that's enough for one day. I say, while we're sitting around, I don't suppose you know how to play cribbage?"

Einstein's reply was never mentioned at the Imperial Geographical Society.

Shakespeare - The Dyslexic Yeare's

Hey-noony-noony?

"Let's see? Ode to a potato? No, needs to be more lyrical. Ode to a duck? Hmm, rhymes with…no, no, can't use that. Ode to a nose, is such a clever prose! Oh joy, I'm such a clever boy!"

At the age of twenty-two, William Shakespeare was still a budding genius. His parents had believed this was true ever since his prize-winning essay on the word 'plonko' had taken him to the head of his class. This was followed by the world's first two-word sonnet and a grocery list done entirely in iambic pentameter. Now he was working on what was to be his greatest masterpiece: a play of Hamlet, Prince of Wapping - 'The story of a moody university student, who refuses to put out the cat, wash his ten year old dreadlocks and murder most foully his mother and uncle.'

Shakespeare was to go down in history, as being the worst and most prolific case of dyslexia known to medical science. Since, however, medical science was not yet a recognized discipline, his parents simply decided he needed some encouragement if he was ever to make it as a butcher. His after-school tutor had been the grave digger who was also the church warden and bell ringer and; his parents believed that anybody who could master that many professions, had a chance of teaching the English language to a boy who routinely mixed grammar with drama in such a way as it turned out porridge.

"Ah, no lad, try not mixing all yee words up again," the Warden would mutter, "like last time, yee can nae be doing that at yee age. And dinnæ chew yee quill."

The lessons had once been conducted in the graveyard on top of a convenient fallen tombstone. Here the church sexton would explain grammatical terms like 'ere by gum' and 'yee doon daft bugger' in such away that young Shakespeare believed they were real. Yet, it was the warden's universal approach to spelling that really turned Shakespeare into a legend.

"Nae, laddy, ye dinnæ spell 'knight' with a k, tis written with a cn, see tis writ cniht nae knight."

To this Shakespeare would cross his eyes and chew on his quill. Then remove all the letters 'k' from his poems and replace them with such ad hoc nouns they would have made more sense in an Albanian sanatorium. Even his cursive writing was more akin to the scrawls of centipedes dipped in ink and left to scurry about the parchment; the boy couldn't spell the dog without his pen managing to explode across the page in a paroxysm of artist interpretation.

"Nae, laddy," the sexton would carefully explain, "dinnæ follow the lines on the paper, just let yee quill fill yon parchments, remember the more yee fill the paper then the less yee ha'e to writ. Writing be nae aboot letters tis all aboot scribbling."

Juxtaposed to the sexton was his actual teacher. A diffident man who lacked the moral courage to simply explain to Shakespeare's parents that the boy simply couldn't spell or - for that matter - read.

At one point his parents had apprenticed him to the village sign writer, but this resulted in the town being renamed from 'Stratford-upon-Avon' to 'Fatsword-Spawn-Anon' and all traffic being diverted away from their shops with accompanying loss of trade.

It was his poetry that was most extraordinary, as this gave him the unbridled freedom to create a new version of the English language in much the same way Picasso moulded clay; having discovered that the key to poetry was not rhyming, but rather words like 'flibbertigibbet', 'gallimaufry' and 'whoobub' that stood out in the listener's mind and whenever in doubt use the phrase 'hey-noony-noony'.

On reaching manhood he moved to London and joined a select group of writers and scribblers called the Lord Chamberlain's Men.

"Let's see," he mused one morning, as his quill danced across the page with the enthusiasm of a mating goose, "oot oot doon spoot! … hmm.. needs more 'ooos'."

At that moment there was the clatter of feet coming up the stairs and the excited cry of Ben Johnson his friend and fellow scribbler.

"Will! Will!"

"Hmm.." Shakespeare kept scratching, ".. to will or not to will. That is the …blast what rhymes … aha … hey-noony-noony..!"

"Will!" Johnson cried at the door.

"Not now! I'm elucidating!"

"You have a commission!"

"Food! At last, we eat!"

"I don't think the money is up front."

"Blast, this means we have to sell off the innkeepers daughter again. Well, what is it?"

"His lord Henry Wriothesley, the Earl of Southampton, has sent this letter, requesting an ode."

"An earl! An ode for the earl! With a wode done in pearl! hey-noony-noony..!" Shakespeare improvised, "Hey, you've opened it."

"I thought there was food inside."

"Inside a thin piece of paper?"

"Wafers or even jam might have been smeared on the paper?"

"You thought the letter was written in jam?"

"Who knows what a lord writes in?"

"But jam?"

"It's been two days since we sold the innkeepers daughter." Johnson held his hands up in supplication. "I am a little vague with hunger."

"Well at least you didn't eat the letter. Oh, I see you have taken a bite out of it."

"I thought there might be jam inside the paper itself."

"Inside? You mean actually … never mind, fetch another daughter of the innkeeper. We need food to be creative. I'm sure we can find someone to sell her off to."

"I think he is out of daughters. Mayhap the wife?"

"She's a bit gammy, isn't she?"

"We could sell the Inn, then?"

"This place hasn't seen a customer since Henry the Fourth died here of the plague. Who in their right mind would buy it?"

"The innkeeper?"

"Brilliant! Oh frabjoyous day!" Shakespeare threw his pen up in the air.

"Is frabjoyous a word?"

"You have so much to learn about poetry."

"I'll go at once."

"Wait, for now is the day of our content."

"Hmm?"

"That means let us enjoy the splendour of the moment."

"Can't we do that after we have eaten? And what moment, we haven't been paid yet."

"Bah! Go philistine, fill thy gaping maw with cram."

"William, just some days, I wish you would use the same language as everyone else uses."

"Art hath no walls, only windows."

"I honestly have no idea what you're on about."

"Go!"

They sold the inn to its owner, who was most bemused to discover the deed was in Shakespeare and Johnson's name, and the fact the legal jargon was bizarrely written in a trochaic of iambic pentameter that squiggled and raced across the paper only impressed him more.

After lunch Johnson and Shakespeare leaned back in their chairs and puffed the new weed from the Americas.

"Cough, ... splutter ... cough .. hack ..."

Then they both threw up their lunch.

"Zounds!" Johnson almost choked, "are we doing this the right way?"

"We must be, it's all the rage you know."

"I mean, I feel so ill, mayhap this isn't tobacco?"

"I bought it from a sailor, it must be tobacco."

Johnson stared at him for a few moments.

"You know," he tempted, "just because you bought it from some sailor, doesn't mean it's from the New World, and it certainly doesn't mean its tobacco."

Shakespeare stared back, and then changed the subject.

"What's the ode supposed to be about?" He asked.

Johnson picked up the half eaten paper.

"Oh, you know, typical fair lady that his Earl is smitten with, or fair boy, one of the two, who is far away and he pines for like a swain."

"Does he say who?"

"Not really. Could be anyone."

"Wasn't Southampton having an affair with one of the queen's ladies in waiting?"

"He was, but they shipped her off to France to hush it up."

"Ah! She is who far away. I bet it's her."

"Maybe, but tis not our problem." Johnson scratched his chin and crushed a louse. "We just need an ode, poem or sonnet to fill up the page. Can't we just use one of your old more legible ones?"

"No! I'm a poet, not a hack," Shakespeare coughed up some more tobacco phlegm and spat on the floor, "besides it won't pay to get caught selling the same shell. Look what happened to Kit Marlowe."

He picked up a pamphlet and read it aloud, "Announcing the Death of Kit Marlowe. Stabbed, then castrated by an unknown assailant, his poems to be investigated."

"Will, you don't think he was killed on account of his rhymes?"

"Yes, why else would someone kill him and then castrate him? That's what I want to know? I mean, he's already dead; it's not like his going to have much use for his codpiece in the hereafter. No, this was the act of a really, really, vicious poetical art critic. It just stands to reason."

"No it doesn't. It doesn't even sit up in bed and beg for a metaphor. He was killed in a drunken bar fight with his own knife." Johnson looked at the pamphlet. "Did you write this?"

"Yes, they paid half a shilling for it."

"So, your whole argument about his being castrated is based on something that you wrote?"

"Yes, but that's not the point."

"That's the entire point, game, set and match!"

"I needed a word that rhymed with investigated!"

Johnson stared at his friend for several minutes.

"William…" he began slowly, "you didn't have anything to do with Kit's death by any chance?"

"How can you suggest such a thing?"

"Well, you are stark raving mad."

"Yes, there is that, but what other reason do you have for giving such offence!?"

"Will! You didn't really have Kit killed, tell me you didn't have him killed?"

"No. He merely died as a result of bad rhyming."

"Hmm. Well, Will, about this ode."

"We need inspiration."

"To the Fount of Wisdom!"

They clattered down the stairs and gave the Innkeeper's wife a good thump on the rump, then danced arm in arm down the street singing:

"Your daughters gonne a dancin' anon.."

"..She hath my heart all gonne awoon.."

"..Sing hey-nonny-nonny!"

"Will, don't you think you could use lines other than a 'hey-nonny-nonny all the time?"

"Oh come now Ben, what's wrong with it?"

"Just about everything, really. It's not even original."

Shakespeare came to clattering standstill and glared angrily at his friend.

"Not original?" He shrilled.

"Hell no," Johnson said back, "Shepherds have been singing 'hey-nonny-nonny' for hundreds of years. Wouldn't be surprised if even the Romans had sung it when they romped about with Boadicea."

"Not … original?" Shakespeare flustered.

"Ah…" Johnson saw the beginning of storm front and he was directly in its path.

"I'll have you know," Shakespeare pointed in Johnson's face, "I spent months coming up with 'hey-noony-noony', and even then it was the result of many, nae endless trips to the fount of inspiration.

Now everyone sings my 'hey-nonny-nonny'. It's a national Anthem!"

"Will," holding up a manuscript for defence, "You heard it in a drunken stupor down at the Fount. Not only is it not original, it's possibly literary piracy."

For a moment it appeared as if Shakespeare would explode as he huffed and mouthed silently, but it passed and in the end he merely said: "Ben, it's too early in the afternoon for critics. Critics belong at the tail end of the dawn, when no one is awake to listen to the putrescent ramblings of deluded minds. Let us to the Fount of Wisdom, ere I forget our friendship."

Johnson fell in beside him, although the tune of their dance was much more subdued.

"Actually, what you just said was rather good."

"Hmm."

"Maybe we should get you angry more often."

"To the Fount!"

"The Fount!"

As they marched in a side door they fell over a drunk.

"You there, fool!" Shakespeare as he lay in a crumpled heap "Why are you disporting yourself upon the lintel?"

"Shakespeare, Johnson, quiet, get down."

"We are down," Johnson pointed out. "You knocked us down."

"Alright, just stay down."

"Who is it?" They whispered back to him. "Why are we lying in a puddle of stale beer and whispering?"

"I'm spying. Now shush. Follow me."

The three of them worked their way through the dark confines of the tavern, knocking over tables and barmaids with a battling average that would have surprised a Welsh cricket team. Eventually, after much 'hey-a-nonny-nonny!" they made it to a table on the far side of the room and crept underneath a broad oaken slab.

"Well, who is it?"

"It's me Marlowe."

"Kit!"

"But you're dead."

"I was," came the cryptic reply, " but for reason of state, her majesty has decided I needed to come back for a short while."

The sheer uncanny nature of the encounter brought the hairs on Shakespeare and Johnson to stand up like pikes on their necks.

"Kit… are you dead or not? This isn't making sense."

"Look, there never was a Kit Marlowe, that's just a name." Marlowe whispered "Killing a name is easy, all you need to do is publicize someone is dead and they seem for all the world to be dead, killing a person on the other hand is a lot more messier and requires some grave robbers."

"I need a drink," muttered Johnson, "Oh look, we're in a tavern."

"Kit," moaned William, "are you dead or not, that's the real question, hey-nonny-nonny?"

"I'm a spy."

"Ooh! I'm so confused." Johnson shook his head. "William did you or did you not write a pamphlet saying Kit was dead?"

"I did, a stranger came to me and said 'here's a shilling, write a story.' So I did. But the whole town knows Kit is dead, the town crier yelled it out, often enough."

"That was me," said Kit, "both times."

"You're the town crier?" asked Shakespeare in amazement.

"And the stranger."

"William, why didn't you recognize Kit when he came to see you?"

"He was wearing a mask."

"So… no, lets just let that one pass on." He looked at Marlowe in the gloom "So you're a spy. I'm beginning to understand. Why are you back?"

"I was never gone."

"I meant why is Kit Marlowe, or whomever you are, back?"

"Because, nobody would expect to see a dead man. It's a disguise."

"So, let me get this right, you're spying by hiding as a corpse in a bar under a table with two of London's greatest poets."

"Almost right," Marlowe nodded, "except for the last part. I'm with three of London's greatest poets."

"Where's the third?"

"That's me you idiot!" Yelled Marlowe.

"Yes, but you're dead!" Riposted Johnson, "You can't be a dead poet!"

"I'm not you blithering idiots! I'm pretending to be dead! What is so hard to understand?"

"It's the bit about hiding under a table while yelling our heads off, that's got me beat," said Shakespeare.

"Alright," whispered Johnson, "so whom are we spying on?"

"The Spanish ambassador." Marlowe said with relief as they finally understood, "He is coming here to have a conversation with two of his countrymen."

"Do you speak Spanish?"

"Damn!" said Marlowe.

"That's a no," Johnson smirked. "I take it."

"I hadn't thought of that," Marlowe grimaced, "damn, damn, damn. Look I don't suppose either of you know how to … no? Damn!"

"I've got it!" cried Shakespeare ecstatically; "lets have a séance with the Spanish ambassador. That way the Spanish can talk to a dead man in English to … Just a thought."

"We could pretend to be Spanish," Johnson prompted, "but speak a very poor dialect, so they have to us speak in English."

"You mean like Irish?" asked Shakespeare.

"Is Ireland part of Spain?" said Johnson.

"They're Catholics," said Shakespeare. "How different can it be?"

"That's it!" cried Marlowe.

"Pretend to be Portuguese?" asked Shakespeare.

"No! Irish Catholics!" Marlowe screamed triumphantly, "The Spanish will talk to them hoping to use them as spies. Brilliant!"

"Yes, I know, hey-a-nonny-nonny."

The head of a barmaid looked under the table and barked at them. "You lot buying or wot?"

They sat themselves at the table and found the entire tavern was silently staring at them.

Marlowe stood up and yelled out "IT'S FOR A PLAY!" Throwing his arms up in a theatrical flourish, this sent all the patrons back to their drinks and games, with a grumbling of how annoying the Arts were becoming.

"It's amazing how effective a disguise being a playwright can be," Marlowe grinned.

"Oh," said Johnson. "Now it's a disguise."

"Look when the Spanish arrive we'll introduce ourselves as merchants from Dublin asking for support from the Spanish crown."

"Alright, but there is one thing that is annoying me."

"Yes?"

"Why do you need us?"

"Because you are known associates of Marlowe."

"But Marlowe is dead," pointed out Johnson not unreasonably, "You killed yourself off."

"Exactly, so people who believe he is dead, will now believe he, or I, am alive again."

"Won't they burn you for a witch?" protested Johnson, "no wait! Won't they burn all of us for being witches?"

"We're no longer living in the Middle Ages, these are illuminated times."

"Yes, but last week didn't they hang, draw and quarter a fellow not two miles from here."

"True, but I was in disguise and it was all a sham. Hark, the ambassador is here."

A short dark man slipped in the door, dressed completely in black, he seemed a silhouette as he minced about the room, deliberately slipping from shadow to shadow to avoid the light. One moment he stood by a pillar, the next swiftly dodging around a bench and beyond a partition to vanish in the dark, then to reappear by a table at the edge of the room. He sat down so quietly and with such poise, it was as if his legs had fallen off and he had deftly landed on the chair. His eyes glittering like stars in the moonlight, watched the room and took it all in with the emotionless stare of an owl on its nightly hunt.

Sir Francis Walsingham was not in fact the Spanish Ambassador; on the contrary he was the very trusted and most dependable spymaster of the English crown. He did, however, slip into these roles with such multifarious ease, that people who had known him for years were sometimes surprised to discover he was a real person.

His life's work was all plots and intrigues, and he was the great master of spies throughout the whole of England with his network of informers ranging from the lowest shepherds of Lake District to the very bedrooms of the Queen's ladies in waiting. He had faithfully served her Majesty with the devotion of a biblical martyr. He did have one fault – old age had withered his mind and he had developed an appalling bad memory, which forced him to keep notes on just about everything. Since the notes were written in codes, which he also often forgot, he occasionally found himself lost in own machinations. For instance, he knew that today, he was supposed to meet someone in this inn, but for the life of him – and in the Elizabethan England the phrase 'for the life of him' often meant exactly that, – he simply couldn't remember whom he was supposed to meet.

He looked up and saw three nervous young men staring at him.

"Would you be mindin' if we are joining you?" Marlowe asked baldly.

"Please do," he pointed at the opposite bench, he felt safe with a solid oak wall behind him. In the back of Walsingham's mind he had the feeling he knew this young man, but couldn't place the face.

A large and somewhat unshaven barmaid appeared before them.

"Wot ya havin?" she barked at Walsingham.

"Ale, please."

"Ale ain't no drink, have a Madeira instead."

"Madeira? Heavens no, I will have a good English ale."

"Out of ale."

"This is an alehouse, isn't it?"

"Ale ain't been delivered yet."

"Very well, cider."

"Out of cider."

"Let me guess, hasn't been delivered either."

"It went off."

"Well, what do you have?"

"We got some Madeira. You can have some of that."

"Damn it woman, the Spanish are our enemies, we can't go drinking their wines. It simply isn't done."

"Do wot ya like matey, but you had better order something or out ya go."

"Very well, just bring some water."

"Dog fell in the well last week, water's poisoned."

"How about some supper instead."

"That's wot killed the dog."

"Gah! Very well, bring me a Madeira, but make it a small one."

"Only have pints."

"A whole pint of Madeira?"

"Coming, right up."

"What? Blast!" said Walsingham, somewhat relieved as the hairy back of the barmaid retreated, "Now then young fellows, how might I attend upon thee?"

"Oh well, you know ..." faltered Shakespeare.

"Yes?" asked Walsingham.

"To be sure, your good sir, we would be like to be havin' some company, ooh ahh ..." recovered Marlowe in what he hoped was an Irish brogue.

"You're French, aren't you?" asked Walsingham.

"Err yes," said Marlowe desperately trying to stay ahead, "I mean Oui."

"Irish!", "Dutch!" Johnson and Shakespeare yelled in unison.

"A mix bag then." Walsingham looked worried. "No wonder you're looking for company. Travellers are you?"

"And Catholic!" Shakespeare almost screamed.

At this point Walsingham felt for his dagger, he may have a faulty memory but he knew whom the enemy was.

"Catholic, you say?"

"Looking for some company," Marlowe almost pleaded.

"Well, you're not Spanish as well by any chance?"

"Are you?"

"I'm not sure I wish to answer that."

"Oh?"

A sliver of a possibility opened in Walsingham's mind.

"Marlowe, is that you?" He leant forward into the candlelight.

"Walsingham!"

"Shush! You idiot. Why are you pretending to foreign Catholics? I hung five of them last week."

"We thought you were the Spanish ambassador."

"Have you ever met the Spanish ambassador?"

"No, but it was dark and you are all dressed in black."

"So are you, you blithering idiot! As are your two friends, are they also the Spanish ambassador?"

"I resent that!" cried Shakespeare, "We are playwrights!"

"So is Marlowe, but I arranged for his official death last week, do you too wish to be a dead poet?"

"This is Will Shakespeare and Ben Johnson," introduced Marlowe.

"Hey-a-nonny-nonny!"

"Oh, you're that Shakespeare." Walsingham sighed. "Maybe I should arrange for your official death."

"Look," argued a frustrated Marlowe, "all I know is the Spanish ambassador is coming here today and I'm here to spy on him."

"Ah," sighed Walsingham, "I remember now, I sent you the letter. Come to think of it, that's why I'm here too."

"Here's ya Madeira." The barmaid returned and slammed the drink down on the top, slopping it over Walsingham, "That'll be a sovereign."

"What?!"

"That'll be a sovereign!" She branded a fist in his face.

"Even a pint of Madeira doesn't cost a gold sovereign!" exploded Walsingham.

"There's an embargo on all Madeira wine from Spain, Madeira is hard to come by, that'll be a sovereign."

"Oh, so there is," Walsingham laughed, "I created it. Ha! Very well, one gold sovereign and well worth it!"

"My lord," began Marlowe, "since it is you, and since we are both here."

"Ah, yes. Very well, let me see," he pulled out a notebook and reading glasses, "Scotland – assassinate the abomination Mary; Sir Francis Drake send to the tower… now have I done that yet? No, never mind, ah ... Spain … Spanish … ambassador ... here it is. 'Meet for ale'. Hmm, that's all, just meet for an ale."

"Are you sure, my lord?"

"No wait, it's in code," muttered Walsingham as he put down his booklet, "Blast, what is 'ale' a cipher for?"

"Zounds! Hey-a-nonny-nonny!" piped up Shakespeare, "let me attend upon this matter."

"You don't speak English by any chance?" asked Walsingham dryly.

"But the matter of the puzzle, is held in the jaw of the muzzle!"

"That's a no, I take it," said Walsingham sarcastically.

"Wait, attend upon my words, for tis plain as curds." The bard continued unperturbed. "Does not 'Meet' rhyme with 'meat' which one doth eat? And doth not 'ale' rhyme with 'ail', whereby a sick man pale?"

"You were doing better with 'hey-a-nonny-nonny', young

man."

"No," Johnson interrupted, "I understand. He means, make ill the Spanish ambassador by poisoning his ale."

"Ah, come to think of it. That may be what I meant. After all, I do order the death of a catholic spy once a week. Why not the Spanish ambassador? My boy, you have the makings of a spy, one of ours I hope, just like Marlowe here."

"Thank you my lord," said Shakespeare, "but I am a bard by profession and do not seek to change."

"My boy, do you really think all that nonsense Kit publishes is poetry? Not on your Nellie! They are state secrets, out in the open for all to see and few to read. And given your background in obscure scribble, I'd say you have the makings of a master, about you."

"Again my lord, I thank thee but..."

"It pays two hundred pounds a year."

"Hey-a-nonny-nonny!" Shakespeare grinned. "Do I get a dagger?"

The barmaid coughed and they discovered she had been standing there the whole time. "That'll be a sovereign."

"Oh what, oh yes," Walsingham said with irritation, "Here, now run off and don't come back in a hurry."

The barmaid stroked her moustache, pocketed the sovereign and then faded into the shadows. What was known to none of them, the barmaid Doris was in fact the actual Spanish ambassador.

Newton, Hell and Nell

love isn't scientific – I. Newton

Newton was in the House of Commons about to present a bill restricting the movement of sheep through the Trinity Great Court at Cambridge University. The shepherds had been taken their sheep on an ancient route through Cambridge. There was nothing the professors could do about this annual intrusion as the shepherds had been given the right of passage by Good King John many years before. Good King John had given this right of way for their help in strangling a pair of troublesome princes who were preventing Good King John from becoming Good King John. Newton in particular was sick and tired of his evening meal being interrupted by a Yorkshire cries of "Git! Git! You's Dozy Dog! Bring it round!", as another shepherd trampled through a flock through Newton's living room.

"Newton," said a voice from out of the clouds as Newton had his head in his notes preparing for his speech, "you really must join me at my gentleman's club for dinner tonight."

Newton turned and looked the newcomer up and down, mostly up. The fellow was a huge florid man whose passions and habits had been well stamped onto his face, his bulbous nose was riddled with red spider veins from a lifetime of over-indulgence of rum and snuff. He towered over Newton like an overripe statue of the god Bacchus.

"Sir, I'm not sure we are acquainted," Newton stared indignantly at an intruder who was so large as to blot out the entire horizon at close hand.

"Philip - Duke of Wharton," the Duke extended a hand that enveloped Newton's like a clam on a hapless diver and refused to let go. "I'm visiting from the House of Lords, wanted to see what you fellows get up to in the lower houses.

"My lord…," Newton began.

"Your Duke," Wharton finished and raised an eyebrow.

"Yes of course my Duke, pleased to meet you," Newton bowed stiffly, not pleased at all and far more concerned about his speech to the commons about unfortunate evils of sheep stampedes in Cambridge.

"Isaac, can I call you Isaac? Of course, I can," the Duke ignored Newton's cringe at being addressed in the first person and continued to maul Newton's hand. "I've heard about your excellent philosophy or is it mathematics. Nevermind, since I've run into you, I'm inviting you to my club to discuss your philosophy with moi. You're the very fellow to bring this philosophy of mine to the lower house and hopefully into some statute of law thing-me-bobbity. After all a lifestyle choice is also a right and has rights before the law."

"My Duke," Newton struggled to release the hand and shook his head incomprehensibly, "I assure you…"

"Good!" Wharton again interrupted Newton, "I will pick you up after your speech thingy in the Commons. Think nothing of it, so happy we have met at last."

"Duke Wharton," Newton finally extricated his hand out of Wharton's punishing grip, "unfortunately I must be returning to Cambridge immediately after I present today's bill. So I must decline your generous offer."

"And your bill is on what precisely?"

"Rampant sheep herding at Cambridge University grounds."

Wharton stared impassively for a moment at Newton then rested his massive hands on Newton's shoulders and bent down to peer into his face.

"Isaac, Isaac, Isaac, I am the chief whip in the House of Lords, trust me I can ensure your bill will be passed by my friends. You do want that - don't you?"

The threat was immediate and palpable. Newton suddenly found himself nodding his head and agreeing to visit the Duke's club despite a terrible feeling he was in the hands of an upper class highwayman. Needless to say Newton's bill was met with uproarious gales of laughter before being passed onto the upper house, the bill was passed partly because Wharton was indeed the Whip of the upper house but more importantly because the members of the King's Loyal opposition were too drunk to vote against it. Afterwards Newton found himself being bundled into a carriage by a man who could have passed for a cross between a dancing bear and the hunchback of Notre Dame, despite Newton crying "Sanctuary! Sanctuary!" from the steps of the House of Commons.

At the house Newton met Nell. Nell was not Lady Wharton as Newton assumed, but rather Nell Gwyn the proprietress of the local brothel which the Duke was calling his club for that night. Nell's face was plastered from wig to ruff in white lead paint to hide the ravages of pox and gin, her teeth were completely replaced with a set of walrus ivory gnashes that proceeded to shoot out every time she laughed. She laughed so often she had taken to tying her teeth to a piece of string and attaching the other end to her bodice to prevent her losing them in a bosom that might have come from the same walrus as her teeth.

"Come in, come in my darlings," she proceeded to bus Duke Wharton on the lips and would have done as well to Newton had he not scuttled backwards in terror from her verminous lips. "Oh ain't he a shy one, my darling, oh I will have fun with this one."

"This is my darling queen Nell," Wharton pushed them together and Newton found himself squirming like a rabbit in a trap.

"Madam!" Newton pushed repugnantly away from the Scylla

and Charybdis of her bosom, "you must excuse me I fear I may be catching something."

"Nothing that I haven't caught already," Nell cackled with a breath that could have melted pewter. "Now then you come in and meet with the children."

"Yes, let's introduce Isaac to our little brood," Wharton remorselessly pushed Newton and Nell before him like a shepherd driving his flock off a cliff, despite Newton's frantic attempts to leap out a window.

"My lord…"

"Do call me Duke," apparently the Duke was used to interrupting, this was something to do with having a hereditary title, "we're all friends here."

"My Duke I thought you wished to discuss your philosophy?"

"I do and we are. We believe, well I believe and everyone does what I say."

"And what pray tell is that?" Newton asked almost not wishing he had to.

"That we should all Do what thou wilt."

"That's it?"

"Do you need more?"

"People can't do what they think they can get away with, the World would be topsy-turvy" Newton protested, "there would be complete anarchy. Supposing the planets did what they wanted to do, Mars would be in Jupiter and Mercury would Venus would always be rising, think of the chaos that would bring to heavens?"

"Think of the fun it would bring to Earth." Wharton open a door and pushed Newton into the room. "We call this the Hellfire Club."

"Mars and Mercury blow me to Aberdeen!" Newton exclaimed, "it's the Devil!"

The room was wall to wall writhing flesh or men in chains; nuns in half habits whipping the men in chains; black candles dripping on black silks; a man dressed as the pope had a trio of young women pouring buckets of wine and milk over him; a quartet of semi-naked musicians played furiously on cellos and oboes; an embroidery of Leda being ravished by a swan encircled the room; as various sheep bleated with their attendant Bo Peeps; Priapic statues satyrs joined with Pan interposed Caryatid columns holding up the ceiling; and in the very centre of it all, was a larger than life statue of Duke Wharton dressed as the Devil having it off with a teapot.

"Precisely," Wharton continued his relentless pushing of Newton into orgy.

"But this is Devil worship!"

Wharton froze in embarrassment and suddenly realized he had chosen precisely the wrong man to champion his ideals to the House of Commons. This took some recalculation and Wharton suddenly saw Newton not a potential all but rather as a potential opponent. He was, however, Duke Wharton the owner of the famous Hellfire club and decided to press on regardless, this inability to retreat from certain disaster was also something to do with having a hereditary title and would later lead some of his descendants to join the Charge of the Light Brigade.

"No," he explained haphazardly, "what I meant is here we mock the Devil. We bring the Devil and his minions here to abuse them and ridicule Hell. For you see we are all good Christians here - What are we everyone?"

"Good Christians!" the revellers chanted in unison and carried on regardless.

"But the whips and the chains?"

"Er ... have you never heard of the Carthusian monks prostrating themselves before our Lord and whipping the Devil out of them?"

"The women are half naked?"

"Um ... was not Eve wholly naked?"

"The huge statue of you doing something obscene with a teapot?"

"Well that's the easiest to explain."

"It is?"

"Must we not mock the Devil with everything we have to hand?" Wharton almost gave up the pretence at this point and waved his hands about in desperation. "After all does not the Devil makes work for idle hands?"

"Yes, but I'm fairly confident this is not how that quote is intended."

"Isaac, Isaac, Isaac ..."

"I really prefer Lord Newton - Warden and Master of the Royal Mint."

"Yes, but I'm a Duke and my preferences come first." Wharton took Newton by the elbow and began guiding him away from the mistake. "Come with me, we can talk in my private chambers."

"I'd rather we didn't talk at all! I must see a man about a ..."

Nell held up Newton by the other elbow and between the two of them they carried him away as his feet scampered on the carpet futilely. In the next room they dropped him on a couch and Nell jumped onto his lap to prevent him from running away.

"Duke I must protest," Newton found himself starting to submerge beneath Nell's overripe body.

"Nonsense," Wharton poured himself a wine and wondered how he was going to get out of this obvious faux pas. "Nell is simply being her welcoming noble self."

"If this is noble I would very much prefer ignoble!"

"Oh Nell does ignoble even better." Wharton laughed,

ensconced himself beside Newton and draped an arm around his shoulder. "Now then."

"Help! Bandits! Fire! Catholics!" Newton cried in despair as he found himself like Daniel with the lions, although in this case the lions were a lot more stranger and his only possible saving intervention might have been actual bandits, fire or Catholics.

"I have brought you here to discuss this philosophy of mine and show you we follow the same mathematical logic that you use to describe the World in all it's motion. The very sort of reason and speculation that Aristotle would have espoused."

"I doubt that very much. If Socrates had done any of this he would have been run out of Athens."

"I have read that you in your inestimable work have shown the World to run as precisely as a clockwork on a church steeple and we can understand the working of the World through logic and reason. I believe that we must apply these same principles of the intellect to understanding religion and faith. For as you said, for every action there is an equal and opposite reaction. So the Lord made the Devil to test our faith, to strength our resolve, to challenge us to bring us closer to our Lord. So we should apply logic and reason to our dealings with the Devil."

"I doubt very much it was the Lord's intention to have a statue of you abusing a teapot as a sign of piety, what's next abusing a Black forest cuckoo clock?"

"Of course not," Wharton looked thoughtfully at the ceiling for a moment and wondered if that was possible, "but my point is it not logical to oppose the Devil by offending the Devil, to mock the Devil and to abuse the Devil? As for every action there is an equal and opposite action? And this is what we do here, we oppose the Devil through our ... acting? We are nothing more than actors in an ancient morality play to educate others to follow the true path to glory."

Newton found himself being won over to Wharton's circuitous argument, but mostly he found himself being crushed beneath Nell's twenty stone of unreasoned body.

"Then who are the audience?"

"We are the spectators! Actors and theatregoers alike! Holding communion with each other for the glory of the Lord."

"Would not proper dress and church going be a far more suitable way of serving our Lord?"

"Would you have us return to the ways of the puritans, would you have us burn witches? Would you have Cromwell and his New Army stamp its terror upon our Isle once more? We here at the Hellfire Club are so very modern, after all do we not have one of the most celebrated mathematicians of our age?"

"Who, sir?"

"You, sir."

"Not I sir?"

"Verily aye, sir."

"Nay, sir."

"Yay, sir, and I'm a duke, sir."

The quartet of semi-naked musicians staggered into the room still playing their cellos and oboes. They had been drinking and playing non-stop for twelve hours and were in desperate need of sleep. They rest of the commune wandered in behind them and the party quickly enveloped the room, followed inexplicably by the sheep who should have had the sense to run away.

Newton realized the only way he was going to get out of this hell was if he agreed for the moment to their terms and then flee at the first possible chance.

"My duke," Newton sighed wearily," what is it you want me to do?"

"Good man," Wharton breathed a sigh of relief, "I want you to sponsor a bill to the Commons enabling the universal acceptance of my Hellfire society throughout the Commonwealth. After all I did just help your bill concerning rampaging sheep in Cambridge pass through the House of Lords."

"What!?" Newton was aghast. "You can't do that!"

"You really do have trouble understanding what a duke can and cannot do." Wharton sank his wine and rose to pour some more, then in mid stride he paused and looked thoughtfully at Nell preening like a Cheshire cat on Newton's lap. "I'll leave you two to get more acquainted."

"Wait? What?" Newton would have leapt up but Nell's lifetime of eating pork pies had ensured immobility.

"I have to see a man about a ...," Wharton voice trailed off as he pranced out the door with the musicians running behind him playing the devil's waltz.

"No, I say, hang on!" Newton struggled as manfully as someone who has spent their entire life doing nothing more manful than pushing an ink pen across a scrap of paper could manage. "For the love of God woman - Get off my lap!"

"Oh dearie," Nell turned Newton's wig back to front and played with his ears, "you're a shy one. Even mister William Wake doesn't carry on like you do."

"Are you telling me you sit on the Archbishop of Canterbury lap?" Newton was aghast.

"Is that what he does," Nell looked surprised, "I didn't think he was a master carpenter then again that would explain the hat, but around here everyone dress a little funny."

"Blast it woman, has the whole world gone mad?"

"Oh dearie he's not the only one, there's mister Walpole."

"The Prime minister?"

"Yes, but he never pays his bills. And there's the fat German fellow, he's such a treat he is."

"You mean George the First of Great Britain?"

"King is he? Who would have thought?" Nell twirled a thread of Newton's wig about her finger. "Must not forget that French fellow. Oh, he's a dresser."

"The French ambassador?"

"No never said he was one of those."

"Don't tell me he's another prime minister."

"No I would know if I had had two of those." Nell looked abstractly in the air. "Louis something, huge nose he has."

"Louis the fourteenth? The Sun King? The greatest ruler in all of Europe?"

"Yes, that's him."

"God in Heaven!" Newton shrank beneath the impossibility of it all. "Is this place the brothel of Babylon?"

"Er now," Nell slapped him playfully with a hand that was mostly mutton and pork, "no need for that fancy talk. I'm a lady I am, that's what my boy Philip says I am, and he's a toff and all."

At that moment Duke Wharton returned to the room with the musicians still struggling to keep not only up with their paymaster but to keep time. There was, however, a third actor that would have brought Newton to his feet, had him running down the hall and have him several blocks away if he wasn't be held firmly in place by a cross-channel juggernaut of sexuality.

"A bear!" Newton found himself hugging Nell even closer.

"This is Felix," Wharton explained as he pulled on a chained and Felix gave a bow.

"It's a bear!"

"For someone who is a world leader in the laws of the planets you do have a tendency to repeat yourself."

"You're got a bear!"

"OK, now you're just getting tedious. I thought you would like to meat our club mascot."

"Why? I mean why are you holding me captive – with a bear! What is this madhouse? What is damnation is going on?"

The Duke poured a gallon of wine into a silver bowl and gave it to the bear. He scratched it behind the ears affectionately and tied its chain to the couch and rejoined Newton and Nell.

"I think you're making too much of a fuss about a silly old bear," Wharton pulled a pipe from his pocket and filled it with a tobacco so rank even the bear wrinkled its nose. "See here Isaac ..."

"Sir Newton."

"Yes, Isaac. The question is not whether the bear is dangerous. The question is not whether the bear wants to tear us limb from limb. The question is whether an Englishman's - or in this case an English Duke's – home is really his castle and can do bally well as he pleases in his castle. I should know I own two castles. And if an Englishman wants to have a dancing bear on his carpet then I say the Englishman has a God given right to have a bear. So what do you say?"

"What in heavens name do you want from me?!"

"I want you to put a bill to the House of Commons allowing universal rights for my Hellfire Club, including a harem and the right to take my bear for a walk in Covent Garden."

"You can't be serious!"

"I'm a Duke it doesn't matter if I'm serious or not, but I do want my bear."

"If I say yes then can I please leave?"

"And a marriage proposal," Nell put in playfully.

"What?!"

"Is that a yes?" Wharton took a taper from the fireside and lit his pipe, then threw the taper over his shoulder.

"If I can leave! Then yes!" spat Newton.

"Another wedding!" Nell gave Newton a slobbering kiss on the cheek.

"Not you woman!"

"And I have your word!" Wharton drew on his pipe and smelled suspiciously at the stench of the tobacco.

"I'm quite certain a word of honour given under torture is not legally binding, so yes damn it all I will put your blasted bill before the Commons, now can you get this blasted woman off of me and let me out of this Bedlam!"

"He gave his word!" Nell shrilled so loudly Newton lost the hearing in one of his ears. "He must be the fourth husband this week. I do like white."

"Right then we are agreed," Wharton pulled a pamphlet from his pocket and handed it to Newton. "Here read this. No Nell there are no monies in it. Is it me or is something burning?"

Suddenly Felix who had been slumbering on his chain gave out a roar of pain and fury as the burning tamper exploded into flame on the bear's fur and the beast reared up and bolted for the door. The couch followed along with it's occupants as Felix screamed in pain and thundered into the next room to the ongoing orgy of wall to wall men in chains and nuns in half habits. The black candles fell over and set fire to the black silks; the man dressed as the pope grabbed a bucket of wine and milk threw it at the bear as trio of young women leapt away quickly followed by the sheep bleating with their Bo Peeps. Bedlam gave way to universal panic as the bear ran into the larger than life statue of Duke Wharton dressed as the Devil having it off with a teapot and knocked it to the ground shattering it completely. In under a minute the club was emptied as everyone spilled out onto the High Street followed by the bear. Duke Wharton fell off the couch and noted to himself it was another successful night at the Hellfire Club.

Newton and Nell were last seen thundering down as Nell was heard to cry: "We're eloping! I'm having a baby!"

Galileo and the Clock

Before the Age of Enlightenment, there was the Age of Darkness.

His children had broken another window in the church again and once more Galileo had to see the archbishop on bended knees.

"But father," he protested, "children are just… well children, you know. These accidents happen."

"This is the third time this month," the bishop chided mildly, "I've been keeping track and I've been keeping the rocks. The rocks are getting larger." He rolled them in his hand.

"Children will grow, your immanence."

"So will the wrath of God, my son."

To this, Galileo had no reply and stared meekly at the floor. After all, there was little he could do anyway, having been under house arrest for ten years since his trial and threat of torture.

"Very well, my son, I have decided you must make restitution for the sins of your children."

"I thought the bible said the children must pay of the sins of their fathers."

"Don't get cocky with me, I could be the Pope one day."

"Yes, your holiness."

"After careful consideration of your history of heretical beliefs on the one hand and your undoubted inventiveness on the other. It seems a waste to hand you over again to the Inquisition simply because of your children's pranks. So, I have decided building a clock in the steeple, where the windows used to be, to be sufficient penitence. Have you anything to say?"

"Not after last time," Galileo muttered, remembering his inquisitional trial.

"What was that?"

"Yes, your holiness," he quickly replied, "I will strive to follow the church's teaching in this …"

"That's not what I asked!" the bishop thumped his chair, "I said build a clock tower – not recant over heresy!"

"Yes, your immanence! Clock tower it is!"

"Much better," the bishop rested his cheek on his hand and looked thoughtfully at Galileo, "now then, tell me – does the Earth move?"

Galileo looked up in surprise, not expecting such a bald interrogation, then sighed in resignation. "No, your immanence the Earth is the centre of the universe, as created by God."

"Good my child," the bishop smiled, "should you ever feel the need to express heretical beliefs then feel free to come to me. I will listen."

Galileo shuddered and swallowed convulsively. "My beliefs are as set down in the holy scriptures, your holiness."

Then another man, who was standing beside the bishop, leaned forward and spoke: "You're a clever man, Galileo," lisped the Grand Inquisitor, "I would hate to burn you."

Galileo Galilei walked away a troubled man. People were trying to kill him. Ordinary, everyday people who had never met him, people with no ambition, scientific drive or even creative ability wanted to stick him up on a pyre of wood and set fire to his underwear. Peasants who worked in fields, joyfully singing songs, having harvest festivals and their beautiful children, all wanted to gouge his eyes out or tie him between angry wild horses just to watch him be torn apart and his intestines spilled out onto the town's piazza. All that, just because the Church didn't agree with he something had said. It was too much. Why did they call it the Mother Church, it was more like a very wicked and spiteful nanny.

Nevertheless, he had the nagging feeling he was absolutely in the right, even if that meant the Fathers of the Church were absolutely in the wrong – this presented a life threatening dilemma. Most natural philosophers had dilemmas like Plato's which went along the lines 'do the gods love the pious because they are pious, or are they pious because they love the gods?' Whereas Galileo's dilemma was not whether the Earth moved around the Sun or vice versa, but could he even mention this question and avoid the flames of the Inquisition?

Not a tall man, his head bobbed up and down on his rotund body as he trundled along in a manner not unlike a wheelbarrow filled with marrows. An odd and almost perverse method of locomotion; was often commented upon by his confessor the bishop as a possible sign of possession by the devil, which lead Galileo to take up jogging, centuries before it became an international craze to lose weight; although for Galileo it was also a way of vanishing from the bishop's sight.

"Excommunicated, excommunicated," he muttered to himself, "if only I was out of communication with the bishop."

At moments like these he would wonder about fleeing northward to Germany or England. He had written numerous letters, all in obscure code, to other philosophers like Descartes, Spinoza and Kepler, but oddly none of them were able to either decode the message or more alarmingly even bothered to reply.

He was clever, he knew that, in fact he was banking on that as being his sole chance of survival – only by using his head and keeping one step ahead of the Inquisition would he be able to keep that head.

His real problem, however, was keeping track of all his children, there were so many they were almost a punishment from

God. Even his wife had trouble remembering all their names. He had tried one year to name them all alphabetically, but this had broken down when he discovered a pile of them left over in the attic which had been left out of his family census, after he had already made one of them a Zappetto. He had discussed with his wife simply tattooing all of them with numbers, but this only resulted in his being banned from the kitchen for a week. The children moved about the house in much the same way as guinea pigs migrated through the Brazilian jungle, squealing, squabbling and the infants leaving mementos in the corners.

Following his trial before the Inquisition, he was, as a good catholic, expected to bring up as many children as his prolific wife's womb could bear. His good friend and confessor the archbishop Piccolomini had told him, that the Inquisition would never deprive newborn children of a father. No, they would wait a year, gouge out his genitalia with red-hot pincers and then burn the heretic. Hence, Galileo Galilei had gone into the children production business with a gusto that alarmed and exhausted his wife.

He arrived back at his home in Siena to a Bedlam. His youngest, or was it his second or third youngest, was crawling down the road following a stray dog, he picked up this one and put it under his arm. Then saw four of his children had been tied to the fence by three more, dressed in bandannas and wafting wooden swords, were taunting their hostages with sweets. On the roof he counted nine, possibly ten, perched chanting a football song for encouragement to the three bandits on the road. His wife was trying to fish one scion out of the well, where the boy was hiding so as not to be tied to the fence.

"Enough!" he yelled and his prodigy scattered like antelope before an enraged lion. Like a lion he had no way of catching but the youngest or infirm, but that was not his intention, he just wanted some peace and quiet.

By the water-well, he reasoned to his wife in a quite rational fashion that filling the well with more water would bring the young troglodyte to the surface. This was met with an icy stare and he was forced to climb down into the well, attach the child to his neck and clamber back up again. It was only when he had retreated to his library did it occur to him the child from the well was actually his neighbours.

He discovered an infant sleeping in his chair, which he carefully removed and stored on the bookcase between Plato and Plutarch. Then sat down at his desk to ponder the problem of the clock tower. It was then he also discovered yet another child had taken up residence beneath the desk and was drawing on Galileo's morning calculations on the orbit of Saturn. He leant down and grunted. The child ran off, calculations in hand, screaming for his mother.

Galileo rested his head in his hands and almost sobbed, and wished he had taken up his father's occupation of making lutes, but Galileo drew himself back determined to not to be defeated by either the Roman Catholic Church or for that matter by his own family, and set to designing the clock.

A clock is no idle tinkering, and a clock tower was a substantial undertaking, a clock tower for the church required the most delicate of social engineering: one mess up and it would be back to a review by the Inquisition. Since, he already knew the interior of the tower, having spent several days of his youth performing experiments on gravity within its solid square walls, he had a fair idea of the size and positioning of the gearing. No, the real problem lay in doing it in such a way, that he got his own back at the dammed bishop without that same bishop realizing it.

He leaned back and saw three of his children, their noses

pressed against the window, staring in at him. The biblical Job with his boils never had such a trial.

Months later he found himself before the tower as it rose above the frozen piazza. It glittered in the frozen day, in much the same way as the Matterhorn had stood before doomed climbers, with that dreadful apex of its summit cutting through the blue sky. Four months had gone by, and miraculously his wife had somehow had given birth to another infant, although Galileo couldn't figure out where she had been hiding it. During that time the Inquisition had not summoned him and the clock for the tower had been finished. All that remained was for him to figure out how to get it in the tower, along with the secret he had openly hidden in its dial.

It had never occurred to him that the bishop would insist on the clock being installed at soon as it was finished.

"But your immanence," he pleaded, "it's the middle of winter, with all the ice it would be very dangerous to send workmen up on roof."

"Workmen?" The bishop glanced at him with a faint smile.

"To install the clock, your holiness," explained Galileo.

"Aren't you doing it?"

"Well, I suffer from vertigo, my bishop."

"The glory of the church comes before your vertigo, my son. You must pray for strength."

"It's not a matter of strength, your immanence, it's a matter of my head colliding with the street below."

"Then you must pray for a sure step and a steady grip. Let's say tomorrow."

"Tomorrow?"

"That's the day after today, preceding a day two days from now," the bishop quipped.

"Tomorrow?"

"You're repeating yourself," the bishop frowned, "you're not suffering from demonic possession again, are you?"

"No, your holiness, tomorrow as the holy church requests."

"Good man, I'll bring a deck chair and lunch to watch the festivities."

The city of Sienna was sport-mad, almost to the point it was a religion in itself and played dangerously close to the boundary of what was acceptable to sharp eye of the Inquisitor and what was heresy. The people needed no encouragement to hold a sporting event; in fact, discouragement or even threats of dire punishments could still manage to create a sense of competition and friendly rivalry that bordered on frenzy. No sport, no competition, not even the occasional bout of warfare with it's neighbouring city of Florence, failed to be accompanied by cheering spectators and enthusiastic flag waving. Even the bishop was known to mutter the occasional hurrah, when one of their teams or even armies took the field.

So, it came as no surprise that whole city had turned out to watch and cheer Galileo as he ascended the scaffolding, after all there was the prospect of him head butting in a home goal with the stones of the piazza. The crowd knew who the teams were, on one side Galileo Galilei, master of the telescope and discovery of the planets and on the other his most reverend the archbishop Ascanio Piccolomini and his forward striker, the Grand Inquisitor.

The Grand Inquisitor was not a kindly man, indeed if he was a kindly man, then the whole definition of 'kindly' is need of some urgent review. The fact that he had no name other than the Grand Inquisitor was enough to give pause to thought to people who never even met him; somewhere along the line, his actual name had disappeared within the Vatican libraries of the eternal city and refused to come out again, that's how unkindly he was. On more than one occasional, the heretics he was to investigate had simply dropped dead from fright when brought before him. His reputation didn't simply precede him, it marched forth as a conquering army setting all to flight and causing all the rest to open their gates and throw themselves upon his 'mercy'. This, however, was an aspect of his personality that was peculiarly missing; presumably it too was lost within the dark confines of the Vatican libraries. He was not a man give much to doubt, which of course made him ideal as an inquisitor, he was of the very strong belief that anyone brought before was 'de facto' guilty of heresy and needed not trial. All that was required was simply a little guidance to return the erring soul to the open arms of a benevolent church, that, or he would burn them to a cinder.

Thus in a wide piazza filled with enthusiastic flag wavers, Galileo was the only one who declined to join in the festivities. Instead, his constricted pupils focused with a monomania – that would have interested Freud and then terrified him – at Galileo as he stood before the scaffolding.

"Signor Galileo," called out the Grand Inquisitor in a soft lisping voice and for once the excited crowd of Sienna fell silent in a way that even the archbishop envied. "Signor Galileo."

Galileo attempted to reply and almost froze with fear before he stammered out. "ye..ye..yes?"

"Good luck."

If this had happened while Galileo had been ascending the scaffolding he would have almost certainly plummeted to his death.

"Thank ... you," he coughed as he gripped a beam, "thank you, your holiness."

The Grand Inquisitor smiled, this made Galileo grip the beam even tighter.

"Don't mention it."

The cheering fans and their flag waving recommenced. It was a cross between a world cup final and an afternoon at the Coliseum with the lions. They didn't really care who was playing or even if they won or not; it was more the whole sense of being involved in something that was bigger than any of them. The fact that someone might die in the process only made it worth talking about afterwards. Galileo's children were also amongst the gathered crowd, their bright red faces blowing steam and hooting encouragement, he hadn't the heart to tell them the danger he faced climbing the icy scaffolding on such a frozen day - and part of him hoped they might cushion his fall.

Galileo turned once more to the scaffolding. It struck him that at the age of seventy-two he was possibly too old to survive this and that also was possibly the intention of the Church's fathers, as it would allow them to wash their hands of the whole matter. This only made him more determined to succeed, or at the very least, bounce when he hit the ground. Success all depended on how cleverly he used this clever head of his. He looked upwards and noticed that far beyond the apex of the tower, the rim of the moon had paused in the sky. He smiled. The irony was not lost on him. Even though the Church had ruled against him and his ideas, the World still sailed on through space and the Moon was another world all in itself.

The clock piece, with its dial, stood next to him wrapped in sackcloth. From this packaged bundle, a rope rose towards a hanging beam and passed back and forth several times over a pulley. He hauled on the rope and this produced a worrying creak, followed by a shower of ice as the strain on the scaffolding sent a snowfall down onto the great scalloped space of the piazza.

"Hurrah!" came the shout of the crowd - but then again the tension was so great they would have cried out if a pigeon had landed in the square. Most people are simple creatures, they may live in houses and carry out conversations, but the houses are designed by cleverer beings and in the entire span of their life most humans never say anything novel. Give them both spectacle followed by monotony and they are content. A chant of "Gal-li-leo! Gal-li-leo!" started up, then quickly subsided as the Grand Inquisitor turned around to stare at them intently.

Hand over the hand Galileo hauled the rope down, and inch-by-inch the clock piece swayed through the air, rising slowly rose up the gantry. Until eventually, a damoclesian quarter ton of wrought iron and precision gearing hung fifty feet above his head. So intense was his concentration he did not notice one of his youngest had crept beneath him and looking up at him from his feet.

"Marina!" he cried out worryingly to his wife.

His spouse, however, was still angry he had given into the bishop in the first place and only returned him a stern silence.

"La bambino! La bambino!" Galileo cried out again.

This was too much for the assembled crowd and they roared as if a goal had been scored in the final moments of an overly tense match, "La Bambino! La Bambino!" thundered about the frozen piazza.

Only then did his annoyed wife feel the obligation to remove the offending mud lark, as Galileo wiped the sweat off his forehead with his shoulder and inched the massive clock piece the final distance up to the summit; you could almost hear Carl Orff's Carmina Burana being played in the background and once more, a tremendous shout went up from the crowd.

He tied off the rope, rested heavily then opened up his fur-lined coat to let out some heat.

"May we see the clock now, master Galileo?" The bishop asked almost politely.

Galileo caught his breath and looked up at the sky, "Soon, your eminence, soon. There is time enough."

"Very droll."

Suddenly he felt a hand on his shoulder and discovered the Grand Inquisitor had crept up on him. "May I aid thee, my son?" came the thin lisping voice of a man who had burned more witches than there actually were in the whole of Italy. Galileo stared in amazement. The last time he had been this close to the Grand Inquisitor the man had been holding a pair of pincers and explaining how they changed shape when heated with fiery coals allowing flesh to be torn out more easily.

"Ye ... yes ... please," Galileo stammered, "could you release the rope when I'm at the top?"

"Of course, my child."

The fact that Galileo was ten years older was not lost on him, nor the possibility that the Grand Inquisitor would be standing beneath a quarter ton of cast iron. He almost absent-mindedly did a rough calculation as to how quickly the clock would descend should he make a mistake, and then did a rough estimate as to how soon his entire family would be arrested and tortured after the all too sudden demise of this man without a name.

He shuddered and began climbing the ladder; his long dark cloaking swirling about him, caught a couple of times on the steps of the lattice ladder. So, he stopped and removed it to make his progress easier. Unfortunately, as he dropped it did in fact fall and land on the Grand Inquisitor. There was a sudden hush, but the Grand Inquisitor drew it off his face and waggled his finger at Galileo and gave him a toothy smirk. In years to come, the people of Sienna would look back to this moment and call it the Miracle of the Cloak – Galileo himself never felt closer to death.

Eventually he made it to the top and almost collapsed in a

frozen puddle of fear. He looked down in the square at the hundreds of people and knew he had made succeeded. It only required for him to slide the clock into the apse of the window and secure it with bolts – anything that happened afterwards was out of his hands. He hooked the ring at the top of the clock piece onto a gantry he had made and signalled down to the Grand Inquisitor to release the rope. At this point he finally found the confidence to remove the sackcloth and stand back to see the response.

There was a wild roar that echoed again and again about the red brick Palio of the city's square, so many silken banners waved in the air it was like the rolling of waves on the Mediterranean.

"Gal-li-leo! Gal-li-leo!" thundered again and again.

He found himself grinning wildly, the world had not collapsed beneath him, it hadn't even open up and devoured his family and he was almost, but not quite, starting to think the bishop was a friendly chap.

The clock face was a shining gold astrolabe of planets, comets and moons, which revolved with the hours. The patron saints of Rome, Sienna and Clockmakers stood as little statues on either side holding up globes and tiny hourglasses. It was a dazzling work of craftsmanship and design, one that would outlast the bishop and the Grand Inquisitor by a thousand years.

Once it had been secured, he made his way once more to the sturdy earth beneath, and giddy with success he amazed he was still able to focus on not falling to his death.

"My son! My son!" The bishop Piccolomini kept saying over and over, as the crowd surged forward and would have carried Galileo about on their shoulders but the bishop forbade this and had his guards force them back.

Amidst all the euphoria Galileo saw the Grand Inquisitor staring thoughtfully at the new clock for a moment, chin in hand,

and then come over almost seeming to divide the crowd like Moses crossing the Red Sea.

"Signor Galileo you have surpassed even yourself," came that thin lisping voice as he shook his hand, "one might almost say – all is forgiven," then he turned and walked off through the crowd.

Galileo let out one last shuddering sigh and looked at his work.

There high, above the square, floated in the midst of the clock face, offset in such a way that only the closet of scrutiny could discern was a model of the solar system where the Sun and not the Earth was at the centre of the World.

www.ingramcontent.com/pod-product-compliance
Lightning Source LLC
Chambersburg PA
CBHW060143260626
47160CB00001B/98